THE CONCISE BOOK OF TIGERS

Outdoorsman, wildlife enthusiast, and nature photographer Lochlainn Seabrook is a popular award winning writer and historian, and the author of 70 books on topics ranging from science and nature to history and religion. Lochlainn does not pen books for fame and fortune, but for the love of writing and sharing his knowledge.

THE CONCISE BOOK OF
TIGERS

A Guide to Nature's Most Remarkable Cats

BY AWARD-WINNING WRITER-NATURALIST AND HISTORIAN

LOCHLAINN SEABROOK

AUTHOR OF THE BESTSELLER *NORTH AMERICA'S AMAZING MAMMALS*

2020

Sea Raven Press, Nashville, Tennessee, USA

THE CONCISE BOOK OF TIGERS

Published by
Sea Raven Press, Cassidy Ravensdale, President
PO Box 1484, Spring Hill, Tennessee 37174-1484 USA
SeaRavenPress.com • searavenpress@gmail.com

Copyright © text and illustrations Lochlainn Seabrook, 2020
in accordance with U.S. and international copyright laws and regulations, as stated and protected under the Berne Union for the Protection of Literary and Artistic Property (Berne Convention), and the Universal Copyright Convention (the UCC). All rights reserved under the Pan-American and International Copyright Conventions.

1ˢᵗ SRP paperback edition, 1ˢᵗ printing, September 2020 • ISBN: 978-1-943737-84-0
1ˢᵗ SRP hardcover edition, 1ˢᵗ printing, September 2020 • ISBN: 978-1-943737-85-7

ISBN: 978-1-943737-84-0 (paperback)
Library of Congress Control Number: 2020944623

This work is the copyrighted intellectual property of Lochlainn Seabrook and has been registered with the Copyright Office at the Library of Congress in Washington, D.C., USA. No part of this work (including text, covers, drawings, photos, illustrations, maps, images, diagrams, etc.), in whole or in part, may be used, reproduced, stored in a retrieval system, or transmitted, in any form or by any means now known or hereafter invented, without written permission from the publisher. The sale, duplication, hire, lending, copying, digitalization, or reproduction of this material, in any manner or form whatsoever, is also prohibited, and is a violation of federal, civil, and digital copyright law, which provides severe civil and criminal penalties for any violations.

The Concise Book of Tigers: A Guide to Nature's Most Remarkable Cats, by Lochlainn Seabrook. Includes illustrations, an index, and bibliographical references.

Front and back cover design and art, book design, layout, and interior art by Lochlainn Seabrook
All images, graphic design, graphic art, and illustrations copyright © Lochlainn Seabrook
All images selected, placed, manipulated, and/or created by Lochlainn Seabrook
Cover photo: "Night Hunting Tiger" © Apiguide

WRITTEN, PUBLISHED, PRINTED, AND MANUFACTURED IN THE UNITED STATES OF AMERICA

DEDICATION

Felis Catus

To the many miniature cousins of *Panthera tigris* who have generously shared their lives with us. You are family.

EPIGRAPH

"...the true tiger is a rare animal, little known to the ancients, and not well described by the moderns. Aristotle makes no mention of the tiger; Pliny only observes of him, that he is an animal of astonishing fleetness, and adds, that he was much more rarely to be met with than the panther, because Augustus first presented a tiger to the Romans at the dedication of the theatre of Marcellus, while, in the days of Scaurus, this aedile presented 150 panthers, and afterwards Pompey exhibited 410, and Augustus 420, at the public spectacles of Rome. But Pliny gives not a single mark by which the tiger is to be distinguished; Oppian and Solinus, who wrote after Pliny, appear to have been the first who mention that the tiger is characterised by long stripes, and the panther by round patches. This is, indeed, one of the marks which distinguish the true tiger not only from the panther, but from several other animals that have been called tigers. Strabo quotes Megasthenus on the subject of the true tiger, who tells us, that, in India, there are tigers twice as large as the lion. Thus the only notices we have from the ancients, concerning this remarkable animal, are, that he is extremely ferocious and fleet; that his body is marked with long stripes; and that he exceeds the lion in magnitude."

Georges Louis Leclerc, Comte de Buffon
1830

CONTENTS

Notes to the Reader - 9
Introduction - 11

1 TIGER BASICS: PART 1 - 13
2 TIGER BASICS: PART 2 - 25
3 TIGER SCIENCE - 37
4 HOME & FAMILY - 51
5 DIET & HUNTING - 57
6 COMMUNICATION - 63
7 FRIENDS & FOES - 67
8 TIGERS IN SUPERSTITION, RELIGION, & LITERATURE - 71
9 VIEWING & IDENTIFYING TIGERS - 81
10 CARING FOR OUR FELINE FRIENDS - 89
11 A TIGER-WATCHER'S GUIDE TO *PANTHERA TIGRIS* - 97

Bibliography & Suggested Reading - 125
Index - 143
Meet the Author - 147

NOTES TO THE READER

PARENTS
☛ Tigers are large meat-eaters. They stalk, kill, and eat other animals for food, consuming them down to their skeletons. Tigers, in turn, are stalked, hunted, and killed by another highly carnivorous creature:  humans, who use tiger parts for everything from folk medicine and rugs to jewelry and meat.

This is the real world of the tiger, a world that encompasses violence in many forms.

While I have done my best to make this book as family-friendly as possible (by, for instance, avoiding unnecessarily graphic images and descriptions), some children may find the contents of this book disturbing. Parental discretion is advised.

ITALICS
☛ Italicized words indicate important facts, concepts, topics, or feline characteristics related to the study and science of tigers.

OLD-FASHIONED
☛ The vintage look and feel of my book is intentional.

LEARN MORE
☛ For those seeking more information or a deeper understanding of *Panthera tigris*, see my bibliography for suggestions concerning further reading.

"There are rich rewards for the man who will enter the lair of the tiger with the purpose of not only finding the inmate of the home, but with a desire to make as many observations as possible which will throw light upon the domestic habits of this royal cat."

Harry R. Caldwell
(1924)

INTRODUCTION

AS A FORMER ZOO KEEPER, and more specifically as a big cat handler who once personally cared for *Panthera tigris altaica*, it is only natural that I would want to write a book about not only one of my favorite animals, but one of the world's favorite animals: the majestic, secretive, charismatic tiger!

The Concise Book of Tigers has come out at an opportune moment in history: I may be one of the last writers to pen such a work, for each year the wild tiger's extinction draws closer and closer. If nothing is done, this magnificent beast will one day become a distant memory, surviving only as a shadow of its former self in artwork, photos, videos, zoos—and books like this one. In the year 3020, a millennium from now, people may very well view it as little more than a supernatural figure in the fairy tales, myths, and legends of the 21st-Century.

Bengal tiger.

For the moment, several thousand wild tigers continue to roam the shadowy grasslands and forests of Eurasia. One can still experience the thrill of seeing a tiger in its native element—proud, untamed, and free. We cannot know how long this amazing opportunity will be available. We do know that time is of the essence.

If this little work not only educates, but also inspires even one of my readers to join in the effort to preserve this beautiful big cat and its ever shrinking habitat, my work will have been a success.

Without question, this enthralling highly endangered mammal deserves our special attention. As the stewards of planet earth—and as the primary cause of this cat's decline to begin with—do we not owe it that much? *Panthera tigris* has been here for at least 2 million years, far longer than we ourselves. In the act of saving it, the wild tiger will repay us and our descendants many times over.

LOCHLAINN SEABROOK
Nashville, Tennessee
September 2020

1
TIGER BASICS - PART 1

INTRODUCTION

Y**OU ARE ABOUT TO ENTER** the hidden jungle world of one of Nature's most intriguing, extraordinary, and fearsome cats: the tiger!

In *The Concise Book of Tigers* you will discover almost everything you need to know about these awesome predators; from where they live, how they care for their young, and what they eat, to how they defend themselves, why they evolved, and the reasons they appear the way they do.

You will get a detailed inside look at the 12 subspecies of tigers, each with an accompanying description that includes its identifying traits, distribution, habitat, feeding and hunting habits, estimated current population, life span, and conservation status—among other important items of interest to feline aficionados.

Tigers are amazing animals and are worthy of study and respect.

As you make your way through the following pages you will learn the hows, whys, and wheres of what I call *tigerology*: how tigers benefit the world ecosystem, why they feature so heavily in literature, art, and religion, where you can go to see them, why wild tigers are disappearing at such an alarming rate, and what you can do as an individual to slow down, and hopefully prevent, their inexorable march toward extinction.

With all of this in mind, let us begin our study of the ecology and life history of the tiger with the word tiger itself.

ETYMOLOGY OF THE WORD TIGER

Like all words, "tiger" has a specific origin and a specific meaning. The modern word tiger derives from French and English, which give us the words *tigre* and *tiger* respectively. These words come from the Latin and Greek word *tigris*, and this derives from the following ancient Persian or Iranian words: *tigr*, meaning "the fast one"; *tighra*, meaning "pointed"; *thigra*, which means "sharp"; and *tighri*, meaning "arrow." All four words are analogous to the Greek word *stizein*, "to tattoo." *Stizein* in turn is related to the Latin word *instigare*, meaning "to kill by piercing, stabbing, or impaling with something pointed."

The original tiger words *tighra* and *tighri* are from an ancient language called Zend or Avestan that was used to write the sacred scriptures of the religion of Zoroastrianism. Zend is one of a group of languages that belongs to the Indo-European language, a family of languages that gave rise to most of the modern languages used today by Europeans (and those of European descent) as well as those living in various parts of Asia.

The names of the Tigris River (shown here) and the tiger share word origins.

Indo-European may date back as far as 9,000 years, to the Mesolithic Period, which also makes it the source for the name of the famous Tigris River in Southwestern Asia: since *tigris* also has the general meaning of "quick" (the river was given this name because it is fast moving), we can see that *tigris* and tiger share common linguistic roots that grew out of the soil of the early Mesopotamian world.

Based on the science of etymology it is apparent that the word tiger evolved from several different but related words, giving it multiple meanings. I have combined these into one definition, which is as follows: a tiger is a *fast*-moving animal that has *pointy tattoo*-like markings and that procures its food by leaping *quickly* upon its prey like an *arrow*, which it then bites with its *dagger*-like fangs, penetrating the flesh with its *needle*-like claws.

This is an apt description of the tiger, illustrating the keen powers of perception and language possessed by ancient and even prehistoric people.

WHAT IS A TIGER?
A tiger is a large four-legged feline that resembles the lion, the jaguar, and the leopard, and which belongs to the same genus: *Panthera*. Tigers are also mammals, a word that comes from the Latin *mamma*, meaning both "mother" and "breast." From this we get the Late Latin word *mammalis*, meaning "of the breast," which was later anglicized to *mammalia*. In taxonomy, Mammalia became the word that designates a class of animals that are warm-blooded, bear live young, have fur or hair, and nurse their babies with breast milk. In English *mammalia* translates, of course, as "mammals."

A tiger is a carnivorous mammal.

A tiger is a *carnivore*, a word that means "flesh-eater." This makes it a member of the taxonomic order Carnivora, the Latin word from which our English word carnivore derives. Though scientists continue to debate the precise definition of "carnivore," many carnivorous animals are characterized by large skulls (in comparison to body size); long sharp teeth; robust bodies; fast twitch (as well as huge endurance) muscles; and strong pointed claws. Thus, this is the category into which the tiger falls.

A biologically precise definition of a tiger then is that it is a *carnivorous mammal*, or more scientifically, a *zoophagous felid*. In plain terms, it is a *warm-blooded, flesh-eating cat*, one that uses its crushing teeth and piercing claws, as well as its tremendous speed and brute strength, to procure its food.

THE COMMON TRAITS OF TIGERS

While there are a number of different types of tigers, in general all are noble looking Asiatic mammals that are noted for their great size, awesome physical power, and absolute fearlessness.

All tigers share most if not all of the following commonalities: large heads; large piercing eyes; small, rounded, erect ears; large but short noses with pinkish-orange nose pads; massive necks; tawny-yellow-gold, burnt orangish, or rufus fawn base fur on the upper body; white base fur on the throat, chest, lower body, and inner legs; a large husky body; four stocky legs; extra muscular front limbs; extra wide front paws; and a long, thinnish whip-like tail.

All tiger subspecies, whatever their size and habitat, share certain basic traits, such as large eyes, small erect ears, vertical stripes, powerful lithe bodies, a long tail, a short nose, and broad front paws.

Except for its nose and chin, a tiger's entire body, from the forehead to the tip of its tail, is covered in irregular black vertical stripes of varying lengths, shapes, and sizes. However, most are dagger- or sword-like in appearance (as mentioned, this is one possible source for its name, as *tighra* means "pointed"). Like a number of other members of the Felidae family, tigers have a bright white spot on the back of each ear.

Each one of these characteristics has one or more important functions, all which we will be exploring in more detail throughout this book.

WORLDWIDE TIGER POPULATION

It is impossible to know the exact number of individual tigers that roam the wilds, but reliable estimates range from 3,000 to 4,000. We also do not know how many captive tigers there are, although an educated guess gives us a range of from 5,000 to 10,000 individuals. However we slice these numbers, they indicate a harsh and foreboding reality: *there are now more captive tigers in the world than there are wild tigers.*

All in all, the total world tiger population lies somewhere between

8,000 and 14,000. It is thought that some 80,000 tigers once inhabited India alone, while at one time there may have been at least 100,000 total worldwide.

NUMBER OF TIGER SPECIES & SUBSPECIES

As is the case with many other animals, there is no scientific consensus as to the number of tiger *species*. While some believe there are multiple species, most maintain there is only one.

When it comes to the topic of tiger *subspecies*, the matter becomes even more divisive and complicated. Some hold that there are as few as two, while others believe there are as many as eleven subspecies; some think there are four, while still others think there are only six. Since not even tiger experts can agree on the exact number, we cannot hope to resolve this issue here.

There are six living tiger subspecies.

Due to the lack of general scientific agreement, in this book I embrace what I consider the most accurate and inclusive model, one that recognizes *one species and 12 subspecies*: six currently living and six that are extinct.

TIGER HYBRIDS

While not recognized as legitimate tiger species or subspecies, we must make mention of two well-known big cat *hybrids*, the *tigon* and the *liger*, since they involve our subject.

A tigon is the offspring of a male tiger and a female lion; a liger is the offspring of a male lion and a female tiger. The tigon and the liger share the physical traits of both species, looking very much like artificial half tiger, half lion creations—which is precisely what they are. Because of this, they are not useful for tiger conservation, and mainly serve as biological curiosities; strange chimeras intended to attract visitors to zoos and circuses.

A less headline-grabbing hybrid, but tragically the most abundant of

Two examples of half tiger, half lion hybrids.

all types of tigers, is the *captive-bred, inter-subspecies, crossed tiger*. Known as the *generic tiger* for short, this is a tiger of unknown genetic background; that is, it was born outside an established tiger lineage, the product of one or more unnamed tiger subspecies. Thus, the generic tiger is distinguished from the *pure tiger*, which is one that is bred within a known and established tiger lineage. This places purebred individuals at the top of the list for conservation, for they have the purest and hence the healthiest genes.

Unfortunately for the posterity of *Panthera tigris*, most pet tigers, as well as many of the captives in zoos and wildlife sanctuaries, are generic tigers. What is "unfortunate" about generic tigers and other tiger hybrids?

Besides sterility, unnatural growth patterns, and various other potentially serious health issues, tiger hybrids possess inferior genes that dilute and weaken the tiger gene pool. Thus, they can neither be used in conservation breeding programs nor returned to the wild.

These are just some of the reasons the practice of interbreeding subspecies—such as mating tigers with lions to serve simply as zoo attractions—has long been heavily discouraged: generic tigers endanger the effort to maintain and preserve pure tiger stock, which is vital if the pure tiger is to survive into the future.

Hybrid tigers are prone to health problems.

While hybrids are still tigers and therefore deserving of humane care and treatment, currently they do not receive the legal protections that pure wild tigers do. There is, however, one area in which generic tigers are provided the same treatment as their more genetically pure cousins: under the Endangered Species Act, hybrids are now subject to identical permitting requirements, a loophole in the law that was closed in 2016. What this means, in essence, is that like purebreds, generic tigers can no longer be sold across state lines without a federal permit. Why is this important?

While this was once legal, one must now obtain an interstate commerce permit or register under the Captive-bred Wildlife Registration program to sell a generic tiger across state lines. This new law will help cut down on the wanton abuse of hybrid tigers, such as subjecting them to overly small and confined enclosures, low quality foods, and poor living conditions—such as a lack of water pools, privacy, soft natural substrate, climbing structures, and environmental enrichment. It will also curtail their unlawful sale, exchange, distribution, and exportation, which when combined is one of the largest illegal commercial trade systems in the world.

TIGER HABITAT

The tiger is both an *oriental native* (that is, it inhabits Asia) and a *palearctic native* (it also inhabits the northern region of the Old World). Having such broad distribution one would expect it to be found in many different biomes and ecosystems, both temperate and tropical—and this assumption would be true.

Tropical coastal thicket.

Indeed, this wide-ranging disbursement explains why there are so many different types of tigers today: over hundreds of thousands of years of evolution, each subspecies became specially adapted to its own unique environment. The Bengal tiger, for instance, tends to live in hot humid areas, hence it has shortish dark fur with dark black stripes—ideal camouflage for damp dusky rainforests; the Amur tiger, on the other hand, lives in cooler climates, and has longer lighter fur with fewer and browner stripes—ideal camouflage for frigid snowy regions with sparser vegetation.

Because *tigers can survive nearly anywhere as long as there is ample food and water*, their habitats cover both low and high elevations (from sea

Mangrove swamp.

level to 13,000'), as well as hot humid climates and cold snowy climates. Thus tiger terrain includes grassland, deciduous forest, plains, scrubland, wetlands, mountains, tundra, pine stands, jungle, marshes, taiga, coastal thicket, alluvial prairie, hill country, rocky regions, rainforest, peat meadows, mangroves, islands, and savanna.

Each subspecies is suited to its own habitat. Whatever the subspecies, however, individual taste often dominates. Some tigers, for example, have a predilection for abandoned villages and old city ruins, while others prefer great expanses of untouched wilderness.

THE TIGER'S RANGE

Tigers once roamed over most of Asia, from Japan on the eastern edge of their range, to Turkey on the western end (on the border of Europe); from Russia on the northern edge of their range, to Indonesia on the southern end.

Today wild tigers survive only in parts of Bangladesh, Russia, India, Bhutan, Thailand, Myanmar, Malaysia, Nepal, China, and Korea (once known as "the land of tigers"). There are indications that tiny populations may still survive in Vietnam and Cambodia. Tigers are now thought to be extinct in Uzbekistan, Tajikistan, Turkmenistan, Kazakhstan, Pakistan, Turkey, Indonesia, Kyrgyzstan, Singapore, Afghanistan, and Iran.

Many Indonesian islands,

Map of Asia, home of *Panthera tigris*.

such as Borneo, do not have tigers. Sri Lanka also lacks tigers, which is probably due to the fact that when the island nation split from mainland India (over 5 million years ago), tigers had not yet evolved.

TIGER TEMPERAMENT & PERSONALITY
An 1807 book summarized the characteristics of *Panthera tigris* like this:

> The tiger is the most rapacious and destructive of all carnivorous animals. Fierce without provocation, and cruel without necessity, its thirst for blood is insatiable; though glutted with slaughter, it continues its carnage, nor ever gives up so long as a single object remains in its sight; flocks and herds fall indiscriminate victims to its fury; it fears neither the sight nor the opposition of man, whom it frequently makes its prey; and it is even said to prefer human flesh to that of any other animal.

While containing a kernel of truth, this sensational 200 year old description utilizes anthropomorphization (ascribing human attributes to something nonhuman)—always a dangerous occupation. Yet, even the most scientific among us must admit that the enigmatic tiger possesses a "humanlike" pride, a "kingly" bearing, and a "devilish" cunning.

This image of the tiger, as a *beautiful monster*, is so deeply embedded in the human psyche, that it has become an archetype; one that haunts not only our dreams and nightmares, but, as we will see, our superstitions, religions, and literature as well.

TIGER SIZES
Of the world's 41 or so cat species, the tiger is the *largest* and *most powerful*, with the biggest tiger growing even more massive than the lion. In addition, the tiger is the *third largest land carnivore* in the world (only the brown bear and the polar bear are bigger).

However, among the various tiger

Tigers come in many sizes and colors.

subspecies, tiger sizes cover a wide spectrum. The *smallest* is the now extinct Bali tiger: the female weighed as little as 150 lbs. and grew to a length of just over 6' (from nose tip to tail tip). The *biggest* tiger subspecies is the Amur tiger: the male can attain an upper weight of almost 1,000 lbs (½ ton) and a length of nearly 14' (from nose tip to tail tip). The other ten tiger subspecies fall somewhere between these two extremes.

A tiger's size can often give us clues as to what type of climate it lives in. Though there are exceptions, according to a popular theory by 19th-Century German biologist Karl Bergmann, generally speaking, animals that live in *cool* or *cold climates* adapt to low temperatures by growing *large thick bodies*. This is because large bodies have a smaller or lower surface area to volume ratio, thus they retain heat better than small bodies. This biological principle, known as "Bergmann's rule," postulates the opposite, as well. Animals that live in *warm* or *hot climates* adapt to high temperatures by growing *small thin bodies*. This is because small bodies have a larger or higher surface area to volume ratio, thus they release heat better than large bodies.

A cool river valley in eastern Russia.

Based on Bergmann's rule, we can see why the Amur tiger, a native of cool climates, is quite large, and why the Indochinese tiger, a native of warm climates, is quite small.

TIGER COATS & PELAGE COLORATION

Tigers have a double-layered fur coat. The hair closest to the body is short, downy, and thick and is called *underfur*. Its main function is to act as an insulating layer, keeping the tiger cool during hot weather and warm during cold weather. Its second coat is made up of long, tough, bristly fur called *outer guard hairs*. This layer acts as a protective shield, a furry "armor" that helps prevent injuries from both plants and other animals. (Many domestic cat breeds have the same double-layered coat.)

Amur tigers in winter, highlighting their magnificent coats.

Tigers *molt* at least once every year, shedding old damaged fur to make way for new stronger hair. As noted, those that live in hot muggy climates usually have short smooth fur and brightly colored coats with dark distinct cross stripes; tigers that live in cool dry climates tend to sport long fluffy fur and are lighter in color with less well defined cross stripes.

All 6 living tiger subspecies have similar coloring: a combination of reddish, yellowish, and orangish base fur with an assortment of *rosettes, spots, and streaks*, as well as *black or brown transverse stripes*, over the head, body, and tail. Tigers are one of the only striped cats, with markings that are unique to each animal, a trait that permits scientists to identify them (just as we can be identified by our fingerprints). In fact, even a tiger's barred side stripes are different on the same individual.

Its stripes, however, are not merely the result of fur coloration, as one might think. A tiger's *skin* is actually striped as well, so that if one were to shave off its coat its stripes would still be visible. No other large felid shares this trait.

What purposes do these furry patterns serve? First, a tiger's stripes are a form of *cryptic coloration*. This is a camouflage strategy in which an animal's colors or markings help it blend into its natural environment, vital to a tiger in both hunting mode and escape mode. Also known as *concealing coloration*, the numerous irregular, vertical, black and orange stripes covering its body mimic the moving shadows typically found in sunlight-dappled forests and windswept grasslands.

Second, its bold striping patterns are a form of *disruptive coloration*, another camouflage strategy, this one designed to blur the normally sharp outline of an animal's body. Though a tiger's stripes are mostly symmetrical on both sides of its face and body, some are asymmetrical, further helping to break up its silhouette against the background.

Tiger coats are also *countershaded*; that is, they are darker on one side and lighter on the other. In the case of the tiger, having darkish fur on the dorsal side or upper body (where there is usually the most light) and whitish fur on the ventral side or lower body (where there is usually the least light) creates an additional camouflaging effect, again, by breaking up the shape of its body while reducing shadows and contours.

Tiger: striped phantom of the forest!

2

TIGER BASICS - PART 2

WHITE TIGERS

CONCERNING COAT COLORATION, OCCASIONALLY A *white tiger* is observed in the wild. This is not an albino nor is it a separate subspecies, but rather a natural genetic mutation, the result of a rare recessive gene carried only by Bengal tigers. In other words, a wild white tiger is a type of Bengal tiger, a genetic variation on the *Panthera tigris tigris* theme. While white tigers can be artificially created (usually to serve as a zoo attraction), this can only be done through inbreeding, which often leads to a host of problems: from birth defects and decreased fertility, to lowered survival rates and increased susceptibility to illnesses.

White tiger.

As for naturally white *wild* tigers, they are quite rare, and for an obvious reason: without its colorful coat and dark stripes, a tiger loses its concealing camouflage. Without this, it also loses the element of surprise, and a tiger that cannot sneak up on its prey will soon die of starvation. Simultaneously, the wild white tiger is also more vulnerable to poachers, hunters, and local villagers. For these reasons the recessive gene that creates a white tiger is rarely passed on to the next generation.

BLACK TIGERS & BLUE TIGERS

Occasionally *black tigers* (which have a dark gray or blackish base coat and greyish stripes) and *blue tigers* (which possess a bluish-gray base coat and dark gray stripes) have been spotted in the wild.

Black tiger.

Like white tigers, neither of these extremely rare felids are true and distinct tiger subspecies. They are the result of what is called *pseudo-melanism*, which is the result of inbreeding (that is, they are the offspring of parents who are closely related)—the black tiger being merely a color variation of the Bengal tiger, the blue tiger (also known as the Maltese tiger) being nothing more than a color variant of the South China tiger.

TIGER LOCOMOTION

Unlike its distant felid cousin the cheetah (who is built for *medium distance high speed running*), a tiger is designed for what I call *short distance burst sprinting*, a strategy that is expedited by several factors:

- Tigers have *small clavicles* or collarbones in relation to body size, permitting both longer strides and less restricted motion when pursuing prey.
- Tigers have *long 3' tails* that help it keep its balance while running, leaping, and wheeling.
- Like most other carnivores, since meat needs to be processed and passed through and out of the body quickly (that is, faster than vegetation), tigers have relatively *short intestines* compared to those of herbivores: the tiger's small digestive system means less weight and bulk to interfere with movement, particularly running.

TIGER SKULL & JAWS

The tiger's large but short spherical skull provides a strong stable platform for its great jaws. This is significant because its jaws must be able to produce a *bite force* capable of taking down large prey animals, puncturing their tough hides, pinning them to the ground, and shearing the flesh from their bones. What kind of bite force is necessary for a tiger to accomplish this?

While, at an astonishing 1,350 pounds per square inch (psi), the jaguar has the most powerful bite force of all cat species, the tiger is not far behind, with a bite force of around 1,050 psi. For comparison the lion has a bite force of 650 psi, the grizzly bear has a bite force of 1,200 psi, and the saltwater crocodile has a bite force of 3,700 psi—making this large reptile the owner of the strongest jaws of any creature in the animal kingdom.

We humans? Our teeth can only clamp down with an average force of about 160 psi.

An important element in the tiger's powerful jaw musculature is its *sagittal crest*, an elevated ridge of bone that runs along the top of the head. Most of us are familiar with the "cone-headed" gorilla. It is its sagittal crest which gives it this curious appearance. The tiger's sagittal crest serves the same function as the gorilla's: it provides extra surface area for anchoring massive jaw muscles.

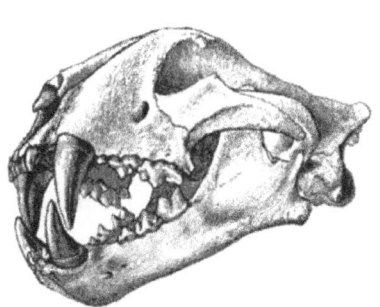
Tiger skull showing the upper jaw, lower jaw, and sagittal crest.

THE TIGER'S FACE
A tiger's *face* is particularly designed to aid it during hunting, while at the same time accentuating its eyesight and hearing:

- *Irregular white and black stripes* over orangish fur cover the face, allowing it to blend into its natural habitat.
- *Large cheek ruffs* (on the male) on each side of the face act like furry radar dishes, scooping up delicate sounds from its environment and funneling them back to the ears.
- *Black and white eye-rings* capture, diffuse, and reflect light (depending on conditions and time of day or night), permitting detailed vision in all kinds of conditions.

Tiger threat face.

When it is excited, in danger, or agitated, a tiger makes a *threatening face*: its pupils dilate, its ears flatten backward, and the sides of its mouth recede, exposing its imposing canines. Anyone who has seen this terrifying mask-like expression will not soon forget it, and this is precisely its function: to instill overwhelming fear. Seeing this horrific visage, its opponent is highly unlikely to attack, and instead will probably either flee or freeze in place. In either case, the tiger will have the advantage.

TIGER EYES

Tigers have *highly specialized eyes* that are perfectly adapted to hunting living prey, particularly at night.

To begin with, unlike domestic cats, who have vertical slit pupils (useful for hunting small prey), tigers have *round pupils* like dogs and humans, which are more useful for hunting large prey.

Second, tigers have *large eyes*, for extra space is needed to house *large pupils* and the many other structures necessary for penetrating night vision. These include *large lenses* (which allow in additional light); *extra rods* (for perceiving motion in the dark); *nictating membranes* or "third eyelids" (which keep the eyes moist and wash away dirt that might impede vision); strategically placed *nerve cells* (for enhanced peripheral vision); and *tapetum lucida* (highly reflective surfaces located behind the retinas, that capture and bounce back even the tiniest amount of light). It is this last structure, a type of mirrorlike tissue, that causes a tiger's eyes to glow at night, an indicator that this is an animal with superlative night vision.

Closeup of tiger eyes and ears.

Foremost is the *location* and *positioning* of a tiger's eyes: like nearly all meat-eaters, they are on the *front of the skull facing forward*. This is quite different than the creatures tigers hunt, for most of their prey, such as deer, wild boar, and zebras, typically have an eye located on each side of the head. How does having frontal positioned eyes benefit tigers?

The scope of each eye converges, resulting in *binocular vision*. This in turn produces a *three dimensional image* that increases depth, detail, and color perception—all four elements which are vital for visually tracking well-camouflaged, fast-moving animals in murky surroundings.

The anatomy as well as the specialized optical structures in its eyes make a tiger's vision six times more accurate than our own.

TIGER EARS & HEARING
Tigers have smallish, erect, rotating ears, aural characteristics that perfectly suit their lifestyle:

- The *diminutive size* of its ears allows the tiger to easily slip through thick foliage and brush quietly and without hindrance.
- The ears' *upright posture* and *top-of-the-head location* provides maximum hearing over long distances.
- When necessary a tiger can *swivel* its ears to the left and right, like little satellite antennae. Pivoting ears allow their owner to pick up tiny sounds in nearly any direction; again, of vital importance when stalking and hunting.

White rear ear spots on circus tigers.

Tigers have *ear spots* or postauricular ear "flashes": a large bright white dot behind each ear. Scientists can only guess at the purpose of these markings, but since a number of other feline species possess them as well, it is obvious that they serve an important evolutionary function, or more likely several.

It is probable that when the spots are "flashed," a tiger is *communicating* anger, fear, excitement, or agitation. There is also the possibility that tiger cubs use them as *homing beacons*, allowing them to more easily follow their mothers through tall dark grass. Postauricular ear flashes may also act as *false eyes*, helping prevent an ambush from behind, for many types of carnivores will only attack if they can do so without being seen. This would mean that a tiger's ear spots serve a second important function as *eye spots* or pseudo eyes.

THE TIGER'S MOUTH & TEETH

A tiger's jaws cannot move from side to side like ours do, or the way the jaws of many other animals do. In fact, a tiger's mouth is designed to open and shut in a *straight vertical line*, an evolutionary adaptation that provides a more powerful bite and a stronger grip. A tiger's dentition and dental arrangement also reveal biological adjustments that suit its lifestyle as a large carnivore:

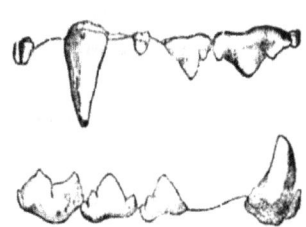

Tiger's teeth showing the large dental gap that allows a deep bite and a firm grip.

- Its small *front incisors* are lined up in a uniform row, allowing these particular teeth to perform detail work, such as striping away hides, plucking feathers, and pulling meat off bones.
- A tiger's *rear teeth*, known as *carnassials*, are built for snipping, cutting, sheering, and slashing flesh.
- Lastly, its *canine teeth*, designed for puncturing and gripping, contain pressure-sensitive nerves, allowing them to "feel out" the best location for delivering a killing bite.

As *terrestrial obligate carnivores*, tigers possess only 30 teeth (we have 32). Compare this to *marine carnivores* (like walruses) who have even less than the tiger: as few as 18 to 24. Yet, *terrestrial omnivorous carnivores* (such as raccoons and bears, which eat both meat and vegetation) usually have 40 to 42 teeth. Let us see why tigers seem to have a tooth "shortage."

A tiger's teeth are not only *extremely sharp*, a tiger is the owner of the *longest*, and some of the *largest*, canines in the feline family, with some measuring up to 4" in length. These, along with its carnassials, are arranged with a *large opening* or space between them, which permits it to both bite deeply into its prey's body and securely restrain it until it has expired. This dental gap is due to the fact that, as just noted, it has only 30 teeth rather than 42. Nature is thrifty and ingenious.

TIGER WHISKERS

A tiger has a wide assortment of *stiff extra sensitive hairs* located over nearly its entire body. Called *vibrissae*, and better known as *whiskers*, they actually function as tactile sensory receptors whose purpose is to detect information about a tiger's physical environment.

Each individual vibrissa or whisker is attached to both a tiger's nervous system and to its musculature, allowing each hair to perceive and move independently. When a tiger's whisker touches an object, a capsule of blood located at its base senses the movement and sends a signal to the brain. This sensorial data helps tigers avoid dangerous situations, find their way through dense vegetation and dark forests, and accurately determine a prey's bite location.

A tiger's facial whiskers.

While most of us are familiar with a cat's facial whiskers, a tiger actually has five different types:

1) *Tylotrich vibrissae*, or body whiskers, are long hairs located over the body.
2) *Carpal vibrissae*, or leg whiskers, are located on the back of the forelimbs.
3) *Mystacial vibrissae*, or snout whiskers, are located around the nose.
4) *Genel vibrissae*, or cheek whiskers, can grow up to 6" in length and are located on the sides of the face.
5) *Superciliary vibrissae*, or forehead whiskers, are located above the eyes.

In addition to a tiger's menagerie of sensitive vibrissae, it also possesses a *special type of skin* on its face. Packed with pressure sensors, a tiger's facial skin can pick up minute changes in the air nearby, a great aid when stalking, hunting, running after, and lunging at its prey. This ability, along with dozens of tactile facial whiskers that can move in any direction, makes its entire head a kind of large sensory tracking device, one from which few animals can escape detection.

A TIGER'S NOSE & SENSE OF SMELL
A tiger is dotted with various types of scent glands over its entire body, from head to tail. It even has scent glands between its toes. This gives each cat a smell that is unique to that individual, one which helps tigers tell one another apart. This is called *olfactory identification*.

For example, tigers regularly rub their bodies against trees, stumps, and bushes, distributing their personal aroma around their territory. Through this form of communication tigers telegraph important information to one another.

Yet, unlike many other carnivorous mammals, the tiger does not rely heavily on its nose to hunt—just one reason it is not as long as a bear's nose or a wolf's nose. In fact, not only does some of a tiger's sense of smell come by way of a structure in its mouth, this structure is designed more for communication than food seeking.

In order to identify the presence of other tigers, for instance, either out of reproductive instinct or territoriality, a tiger uses a chemically-sensitive smelling receptor located in the upper part of its mouth. Called the "Jacobson's organ," it allows a

Tiger displaying the flehmen response.

form of mammalian communication known as the *flehmen response* (or simply flehmen), in which an animal curls its upper lip (baring its teeth in a silent snarl), lifts its head, crinkles its nose, and inhales in response to olfactory stimuli.

The flehmen position allows air to more easily transmit scents (such as pheromones) to the Jacobson's organ. While this pocket-like chemoreceptor is found in amphibians, reptiles, and mammals, the flehmen lip-curl response itself is a strictly mammalian behavior, and is a familiar behavior to owners of horses, llamas, goats, and domestic dogs and cats.

As we have seen, its nose is short because the tiger relies on other senses more than smell. Yet, its smallish size actually serves an important purpose: a *short nose* allows extra room for the growth of thicker skull bones, which, in turn, forces the brain to grow smaller and more compressed. Both of these characteristics further protect and strengthen the tiger (and its head region) against the many dangers and stresses it faces as a large apex predator.

A TIGER'S TONGUE & SENSE OF TASTE

Those who own *Felis catus* (the domestic cat) are familiar with their rough tongues; and their gargantuan cousin, *Panthera tigris*, is no exception to this rule. A tiger's tongue too is covered in backward-facing, sandpaper-like, fleshy bumps called *papillae*—better known as "taste buds."

The grating surface of these tiny sensory organs is a felid adaptation that allows tigers to efficiently remove meat from bones and feathers from skin. Just as importantly, a tiger uses its tongue for *grooming* its luxuriant coat: the sharp raspy taste buds are ideal for getting rid of old matted hair and loose fur, for dislodging dirt, and for dispersing glandular oils over the skin and pelage.

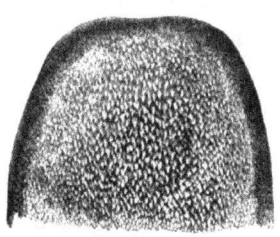
A tiger's raspy tongue.

A tiger's *salvia* contains powerful antiseptic chemicals, which it spreads on cuts and other injuries with its tongue. This not only helps mend its wounds, but it also aids in the prevention of infections while they are healing.

Like humans, tigers are able to distinguish between basic *flavors*, such as sweet, salty, and sour. Yet, despite its large specialized tongue, a tiger has far less taste buds than we do, an indicator that it is not an especially finicky eater and that its sense of taste is not of vital importance in its day-to-day life.

TIGER LEGS, FEET, & CLAWS

Tigers have four *powerful legs*. The forelimbs come specially equipped with heavier bones and bigger muscles than the rear legs, an advantage when it comes to knocking down and pinning large animals. Its hind legs, however, are lighter and longer than its front legs, an adaptation that provides extra jumping power for pouncing on its food. When necessary a tiger can leap a distance of nearly 35'.

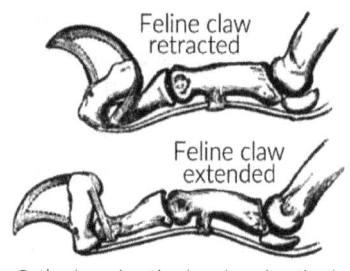

Cat's claw sheathed and unsheathed.

A tiger's *paws* are heavily padded, enabling it to walk long distances and run short distances while generating little or no noise. In addition, its *foot bones* are tightly wound with tough but flexible tissues called *ligaments*, which serve several important functions: these fibrous cords make its feet stronger and thus better able to absorb the tremendous forces the cat experiences when running, jumping, and landing. Thus its fleshy reenforced paws not only act as *sound dampeners*, but also as *shock absorbers*, allowing it to stalk, hunt, and leap, both without injury and in near complete silence.

A tiger is a *digitigrade* animal, that is, it walks on its toes rather than on the soles of its feet. (In contrast, creatures that walk on their soles, like we do, are called *plantigrades*, while those that walk on their hooves, such as horses and deer, are known as *unguligrades*.)

While walking, tigers use what is called a *pacing gait*: they lift both legs on the same side. After these are placed down, the two legs on the other side come up off the ground. Therefore, while walking, a tiger always has two legs in the air. (The tiger's tiny cousin, the domestic cat, also employs the pacing gait.)

Fast-running cheetahs have semi-retractable claws that are suited to their lifestyle and hunting habits.

Like most cats large and small, tigers have *fully retractable claws* (cheetahs, an exception, have semi-retractable claws). In the tiger's case, it has four fully retractable claws on each paw, all which reach gigantic sizes, in some subspecies (such as the Bengal tiger) up to 4" in length. When relaxed or when not needed, ligaments draw these fearsome weapons up into special skin sheaths in order to protect them from becoming dulled on hard surfaces, such as rocks and desert soil. In other words, retraction helps keep claws razor-sharp for when they are needed most: during defense and hunting. Incidentally, the extension of a tiger's claw has been likened to the unsheathing of an Indonesian karambit, a type of curved claw-like knife.

Being heavily *curved* and *pointed*, tiger claws are ideally designed for grabbing and immobilizing prey, as well as for climbing hills, rocky outcrops, and trees. Unlike some other cats, however, tigers cannot go down trees headfirst. They must climb awkwardly down backwards or leap to the ground.

A fifth claw, called a *dewclaw*, is located about where a human thumb would be; and in fact, dewclaws act very much like our thumbs, giving tigers an edge when it comes to grasping, holding, and climbing.

TIGER SENSES
While, as we have seen, eyesight, taste, smell, and touch are important to tigers, by far their most sophisticated mode of perception is *hearing*. As it is a carnivore whose survival depends on hunting and killing other creatures, it is not surprising that a tiger's aural faculty is its most highly developed sense. What may astonish many is just how sensitive a tiger's hearing is.

Tigers can hear both *infrasounds*, low pitched frequencies below what we can hear, and *ultrasounds*, high pitched frequencies above what we can hear. Here is some perspective.

Normally we hear sound frequencies from roughly 20 Hz to 20,000 Hz, a spectrum known as the "acoustic range." Sounds *below* the acoustic range (that is, below 20 Hz) are in the infrasonic range; sounds *above* the acoustic range (that is, above 20,000 Hz) are in the ultrasonic range. Tigers can easily hear sound frequencies below 20 Hz and probably up to as high as 200,000 Hz. In reality, its sense of hearing is so great and so complex that the lowest and highest ends of the tiger's auditory capabilities have yet to be precisely determined or described by science.

Tigers possess acute physical senses, especially hearing.

TIGER COMMUNICATION

Chapter 6 is entirely devoted to this topic, so let us move on to a tiger's life expectancy.

TIGER LIFE SPAN

Although wild tigers usually live around 10 years on average, 15 years is not unheard of. Captive tigers can live for over 20 years. These statistics are for adults. In comparison, tiger cubs in the wild do not fare as well: at least 50 percent of all wild tiger babies die before they are two years of age.

If this tiger cub survives adolescence it may live for between 10 and 20 years.

A rough estimate of a tiger's age can sometimes be gauged by various physical signs and characteristics. Broken and missing teeth, for example, along with dull stripes and a faded or unkempt coat, may indicate an aged individual. On the other hand, an energetic tiger with bright coloration, well maintained pelage, and a full set of teeth is more likely to be a teenager or young adult.

A hunting party cornering a wounded tiger.

3

TIGER SCIENCE

TIGRIPHILIA

IF YOU LOVE TIGERS THEN you have what I call *tigriphilia* and you are a *tigriphile* (one who has a "love of tigers"). This "condition" is also sometimes known as ailurophilia (a "love of cats"). Of course, there are people who dislike and fear tigers. These individuals possess what I have termed *tigriphobia*, making them *tigriphobes* (one with a "fear of tigers").

Whether you are a tigriphile or a tigriphobe, you will not only find the science behind *Panthera tigris* informative, but fascinating as well.

THE CLASSIFICATION OF TIGERS

Like humans, tigers are arranged on a "family tree" according to *taxonomy*: the science of classifying animals and plants in orderly categories, based on their evolutionary relationships.

To accurately classify animals, a number of categories have been created. The more common ones are as follows:

Tigers are classified according to taxonomy.

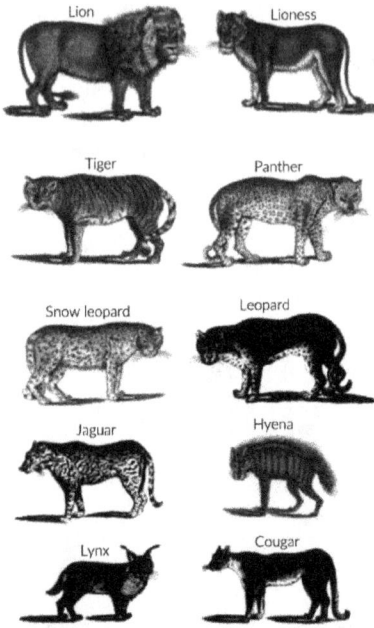

Members of the suborder Feliformia.

Superkingdom, Kingdom, Subkingdom, Infrakingdom, Phylum, Subphylum, Infraphylum, Super Class, Class, Subclass, Infraclass, Order, Suborder, Family, Subfamily, Genus, Species, and, when necessary, Subspecies.

Taxonomists (scientists who classify animals) put each of the many different tigers into one family, which, like all scientific names, is written in Latin or Greek: Felidae, that is, "cats."

As discussed earlier, most, but not all, biologists seem to agree that there is only one tiger species, which goes by the binomial: *Panthera tigris*. However, there is no scientific agreement on the number of tiger subspecies. Thus, I have chosen what I consider the most intuitively logical model: one species, *Panthera tigris*, and 12 subspecies—six living, six extinct.

The 12 tiger subspecies are categorized *phenotypically*; that is, they are arranged according to various observable traits that have been defined by the environments they inhabit and by the makeup of their genes. *Phenotypes* include such things as behaviors, blood type, skull shape, body size, and coloration.

WILD CATS WORLDWIDE

Tigers are just one member of a large, noble, prehistoric group of wild cat species. These number between 36 and 42, depending on which authority you choose to accept. Why the discrepancies?

No one can agree on how to clearly and scientifically define cats as a group, let alone a single distinct species. Thus, we must be content with my personally selected list of 41 wild felines, which, in alphabetical order, looks like this:

Key: Common Name/Scientific Name
1. African golden cat: *Caracal aurata.*
2. African wild cat: *Felis lybica.*
3. Andean cat: *Leopardus jacobita.*
4. Asiatic golden cat: *Catopuma temminckii.*
5. Black-footed cat: *Felis nigripes.*
6. Bobcat: *Lynx rufus.*
7. Borneo bay cat: *Catopuma badia.*
8. Canada lynx: *Lynx canadensis.*
9. Caracal: *Caracal caracal.*
10. Cheetah: *Acinonyx jubatus.*
11. Chinese mountain cat: *Felis bieti.*
12. Clouded leopard: *Neofelis nebulosa.*
13. Domestic cat: *Felis catus.*
14. Eurasian lynx: *Lynx lynx.*
15. European wildcat: *Felis silvestris.*
16. Fishing cat: *Prionailurus viverrinus.*
17. Flat-headed cat: *Prionailurus planiceps.*
18. Geoffroy's cat: *Leopardus geoffroyi.*
19. Guiña: *Leopardus guigna.*
20. Iberian lynx: *Lynx pardinus.*
21. Jaguar: *Panthera onca.*
22. Jaguarundi: *Herpailurus yagouaroundi.*
23. Jungle cat: *Felis chaus.*
24. Leopard: *Panthera pardus.*
25. Lion: *Panthera leo.*
26. Mainland leopard cat: *Prionailurus bengalensis.*
27. Marbled cat: *Pardofelis marmorata.*
28. Margay: *Leopardus wiedii.*
29. Mountain lion: *Puma concolor.*
30. Northern oncilla: *Leopardus tigrinus.*
31. Ocelot: *Leopardus pardalis.*
32. Pallas' cat: *Otocolobus manul.*
33. Pampas cat: *Leopardus colocolo.*
34. Rusty-spotted cat: *Prionailurus rubiginosus.*
35. Sand cat: *Felis margarita.*
36. Serval: *Leptailurus serval.*

Leopard.

Pampas cat.

Lynx.

37. Snow leopard: *Panthera uncia*.
38. Southern oncilla: *Leopardus guttulus*.
39. Sunda clouded leopard: *Neofelis diardi*.
40. Sunda leopard cat: *Prionailurus javanensis*.
41. Tiger: *Panthera tigris*.

THE BIG CAT FAMILY

Bengal tiger.

Despite being categorized with 40 other cats, the tiger possesses at least one primary difference; one that sets it apart from all other members of the Felid family. I am speaking of its *great size*, for it is the largest living cat in the world. It is because of this that the tiger has been labeled a "big cat" and placed in the Felidae subfamily known as Pantherinae, a group that represents the *Panthera* lineage as a whole.

Besides the tiger, the *Panthera* lineage—which is itself further divided into two genera (*Panthera* and *Neofelis*)—includes six other big cats. Here is a complete alphabetized list of all seven species:

1. Clouded leopard (*Neofelis nebulosa*).
2. Jaguar (*Panthera onca*).
3. Leopard (*Panthera pardus*).
4. Lion (*Panthera leo*).
5. Snow leopard (*Panthera uncia*).
6. Sunda clouded leopard (*Neofelis diardi*).
7. Tiger (*Panthera tigris*).

Jaguar.

THE TIGER'S FAMILY TREE

A true scientific understanding of the tiger can only be grasped by seeing how and where it fits in the *Web of Life*: the global ecological community of all lifeforms (including humans), a biological system that makes it easier for us to see how earth's many organisms are related and interconnected. Systematists, specialists in the field of taxonomy, categorize tigers like this:

- The Superkingdom or Domain to which all tigers belong is *Eukarya* (meaning "true kernel"): eukaryotes are organisms whose cells have a nucleus.
- The Kingdom to which all tigers belong is *Animalia* (meaning "animals"): animals are living creatures that have the capacity to move and respond to stimuli quickly.
- The Subkingdom to which all tigers belong is *Bilateria* (meaning "bilateral"): bilaterians are living creatures that are symmetrical (identical) on the right and left sides of their bodies from head to tail.
- The Infrakingdom to which all tigers belong is *Deuterostomia* (meaning "second mouth"): deuterostomes are living creatures who, when embryos, first develop an anus and secondly a mouth.
- The Phylum to which all tigers belong is *Chordata* (meaning "cord"): chordates are animals that have a bundle or "cord" of nerves running down their back.
- The Subphylum to which all tigers belong is *Vertebrata* (meaning "jointed"): vertebrates are animals that have a spinal column with a cord of nerves inside.
- The Infraphylum to which all tigers belong is *Gnathostomata* (meaning "jaw mouth"): gnathostomes are animals with jaws.
- The Superclass to which all tigers belong is *Tetrapoda* (meaning "four feet"): tetrapods are animals with four limbs.
- The Class to which all tigers belong is *Mammalia* (meaning "mammals"): a mammal is a warm-blooded animal with a spine and a cord of nerves, whose body is covered with fur, who gives birth to live young, and who nurses its offspring with milk.
- The Subclass to which all tigers belong is *Theria* ("beast"): therians are a group that includes both placental and marsupial mammals.
- The Infraclass to which all tigers belong is *Eutheria* ("true beast"): eutherians are placental mammals whose young are nurtured inside the body (as opposed to marsupial mammals whose young are nurtured outside the body).
- The Order to which all tigers belong is *Carnivora* (meaning

Vertebrae.

- "carnivore"): carnivores are meat-eaters that must consume flesh to survive.
- The Suborder to which all tigers belong is *Feliformia* ("having a cat-like form"): feliforms only have to possess the general appearance of a cat, and so some feliforms are not actually cats, such as, for example, the hyena. (The Feliformia suborder is also sometimes written Feloidea.)
- The Family to which all tigers belong is *Felidae* ("resembling cats"): felids are cats that are characterized by powerful muscles, large chests, strong forelimbs, and a predatory lifestyle.
- The Subfamily to which all tigers belong is *Pantherinae* ("pertaining to panthers," or more loosely, "concerning big cats"): pantherines are cats that resemble panthers, that is, large cats. The word panther itself derives from the Greek words *pan* ("all") and *ther* ("beast"); in other words, pantherines are the archetypal beasts or nonhuman animals.
- The Genus of all tigers is *Panthera* ("panther," that is, a leopard, or more loosely, a "big cat"): the pantherids are a group containing the four so-called "roaring cats"—tigers (*Panthera tigris*), lions (*Panthera leo*), jaguars (*Panthera onca*), and leopards (*Panthera pardus*).

Black panther.

- The Species to which all tigers belong is *Panthera tigris* (meaning "panther-tiger"): a group composed of large, powerful, orangish cats with dark stripes that resemble panthers.
- As noted, all tigers are broken into Subspecies: a biological division below the species level that is made up of genetically distinguishable and geographically separate populations, but which are capable of interbreeding. The current tiger subspecies are (according to my personal view):

    1. Amur tiger (*Panthera tigris altaica*): living.
    2. Bali tiger (*Panthera tigris balica*): extinct.

3. Bengal tiger (*Panthera tigris tigris*): living.
4. Caspian tiger (*Panthera tigris virgata*): extinct.
5. Indochinese tiger (*Panthera tigris corbetti*): living.
6. Javan tiger (*Panthera tigris sondaica*): extinct.
7. Malayan tiger (*Panthera tigris jacksoni*): living.
8. Ngandong tiger (*Panthera tigris soloensis*): extinct.
9. South China tiger (*Panthera tigris amoyensis*): living.
10. Sumatran tiger (*Panthera tigris sumatrae*): living.
11. Trinil tiger (*Panthera tigris trinilensis*): extinct.
12. Wanhsien tiger (*Panthera tigris acutidens*): extinct.

A TAXONOMIC EXAMPLE

To further our knowledge of the tigers' family tree and how scientific classification works, let us now look at a single example, the Amur tiger. Here is its individual taxonomic family tree:

Superkingdom: *Eukarya*.
Kingdom: *Animalia*.
Subkingdom: *Bilateria*.
Infrakingdom: *Deuterostomia*.
Phylum: *Chordata*.
Subphylum: *Vertebrata*.
Infraphylum: *Gnathostomata*.
Super Class: *Tetrapoda*.
Class: *Mammalia*.
Subclass: *Theria*.
Infraclass: *Eutheria*.
Order: *Carnivora*.
Suborder: *Feliformia*.
Family: *Felidae*.
Subfamily: *Pantherinae*.
Genus: *Panthera*.
Species: *Panthera tigris*.
Subspecies: *Panthera tigris altaica*.
Common name: Amur tiger.
Alternate common name: Siberian tiger.

Amur tiger.

THE IMPORTANCE OF SCIENTIFIC NAMES

When a *mammalogist* (a scientist who studies mammals) wants to identify a particular tiger, he or she does not use the tiger's entire family tree, of course. It is far too long. And besides, every category up to "Species" is the same for every type of tiger. So none of the categories before this particular one need to be written out.

Margay, *Leopardus wiedii*.

How then do scientists identify a specific tiger (or plant, fish, or bird)? They use what is called *binomial nomenclature*: binomial means "having two names"; nomenclature means "a standardized system of naming things." In binomial nomenclature only the last two categories, the genus and the species, are used for identification. These two categories are combined into one name—beginning with the genus and ending with the species—and are written in italics. (The name of the genus is always capitalized; the name of the species is never capitalized.)

When dealing with tigers, however, we must use what is known as trinomial nomenclature: a naming system that uses three names. This is because all of the known tigers are subspecies, which requires the use of three names: the genus name, the species name, and the subspecies name (again, all which are italicized).

With this knowledge in mind, let us go back to the Amur tiger's family tree. How would the Amur tiger's scientific name be written using binomial nomenclature? The answer is *Panthera tigris*. How would it be written using trinomial nomenclature? *Panthera tigris altaica*.

Liger (lion-tigress cross).

There is another important reason scientists use the genus, species, and, when necessary, subspecies names when they want to talk about a specific tiger. Common names, like "Amur tiger," can, and do, change over time and from region to region. Depending on the country, for example, the Bengal tiger (*Panthera tigris tigris*) is also known as the "Indian tiger," the "White tiger," the "White Bengal tiger," and the "Royal Bengal tiger." From this example alone one can see how common names can lead to confusion.

To prevent this type of misunderstanding, scientists use an animal's scientific name, which is always written in Latin or ancient Greek; or a combination of the two. Why? Because Latin and ancient Greek are *dead languages*, meaning that they are no longer used in everyday speech anywhere in the world (in contrast, *living languages*, like English, are constantly changing and expanding).

As a result, Latin and ancient Greek, like all dead languages, never vary, no matter how much time passes, and no matter what part of the world you may be in. In this way, the scientific (that is, Latin-Greek) name remains the same over time, always making precise identification of tigers (or any other type of animal, plant, or mineral) possible.

Lion.

Despite this seemingly ironclad fact, there are exceptions. For example, an animal's scientific name can and does change on occasion, such as when new discoveries about its ecology and life history demand that it be renamed or placed in a new category. In fact, this has already occurred several times in the tiger world. The most recent change in tiger taxonomy came in 2017, when the International Union for Conservation of Nature (the IUCN) revised the list of tiger subspecies. As our knowledge grows and technology becomes more sophisticated, continued revisions such as this are to be expected in the future, and, by the time you read this book, may have already occurred.

THE TIGER'S CLOSEST LIVING RELATIVE
Contrary to popular assumption, the tiger's nearest cousin is not the lion (*Panthera leo*), the jaguar (*Panthera onca*), or even the leopard (*Panthera pardus*). It is the snow leopard (*Panthera uncia*), who, besides having a close genetic kinship, has several things in common with its larger striped relative, including some striking parallels in morphology (physical form and structure).

Snow leopard.

Known as "the ghost of the mountains" due to its secretive nature, there may be as few as 5,000 snow leopards left in the wild. The precise population of this mysterious rosette-covered cat is impossible to determine, for its range covers at least 12 countries and massive tracts of remote wilderness scattered over central Asia. What is beyond doubt is that, like the tiger, it is a rare and extinction-vulnerable feline that needs our protection.

TIGER EVOLUTION

The precise manner in which tigers evolved is uncertain, and even what little we do know about their *phylogeny* (evolutionary history) is unsettled and vigorously debated. Even the first cat, that is, the ancestor of the entire felid family, remains lost in the shadowy mists of prehistory. *Cat-like animals* appear in the fossil record between 60 and 50 million years ago, while *true cats* seem to have emerged about 38 to 35 million years ago.

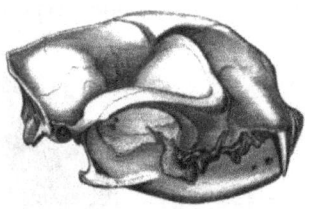

A primitive cat skull, *Archelurus debilis*.

One popular theoretical family tree designates a group called the *Miacids* as the earliest carnivorous mammals. According to this view, the Miacids gave rise not only to dogs and bears, but also to Viverravidae, which produced the cat line known as Feloidea. About 30 million years ago the Feloidea then gave rise to a primitive cat known as *Proailurus*, which could be the last common cat ancestor, making it the progenitor of all felids.

This small 2' long feline probably went extinct some 23 to 20 million years ago. Proailurus, whose name means "before cats," may have looked something like a

Saber-toothed tiger (not a true tiger).

prehistoric cross between a domestic cat, a hyena, a civet, and a mongoose. In fact, all four animals are members of the suborder Feliformia, and are thus related.

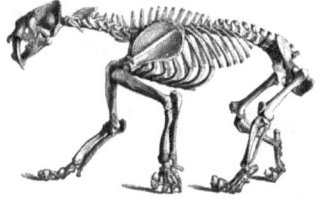
Saber-tooth cat skeleton.

Before completely disappearing, however, during the Miocene period the Proailurus genus gave birth to both the now extinct saber-toothed cat line, *Machairodontinae*, and *Pseudaelurus*, a prehistoric cat that flourished in Asia, Europe, and North America around 20 million years ago.

Around 18 million years ago, Pseudaelurus produced *Schizailurus*, the probable ancestor of modern cats. According to this view, it was this particular feline which gave rise to the modern living cat line, Felidae, the family we are most interested in. For it is the Felidae group that contains all living cats, including the 3 million year old genus *Panthera*, whose members are tigers, lions, jaguars, leopards, and snow leopards.

Our present-day understanding of the emergence of tigers as a distinct species is based on the currently known oldest tiger fossils, which were discovered in China—making the tiger a native of eastern Asia. These particular remains have been dated to roughly the beginning of the Pleistocene Epoch, between 1.8 and 1.6 million years ago. By 12,000 years ago tigers had spread into what is now India. The precise time of their arrival in eastern Russia is unknown, but it was sometime during the Pleistocene (from 2.6 million years ago to about 11,700 years ago).

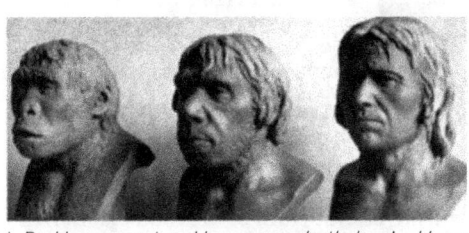
L-R: *Homo erectus, Homo neanderthalensis, Homo sapiens*. The human species on the far left, *H. erectus*, lived in Eurasia during the time primitive tigers first appear in the fossil record, and no doubt competed with them for food, territory, and other natural resources.

All 12 of the currently known tiger subspecies seem to have come from a prehistoric ancestor that lived around 100,000 years ago or so, and which may have looked and behaved something like the living Indochinese tiger. Another theory maintains that, due to its many primitive features, the South China tiger may be the forerunner of all 12 tiger subspecies.

There is still much to learn about the origins and evolution of *Panthera tigris*.

WHY ARE THERE TIGERS?

According to mainstream science, tigers were created by *evolution*. According to mainstream religion, tigers were created by *God*. Which ever explanation one chooses to embrace, it is clear that tigers—whether here accidently or intentionally—play a vital role in the world ecosystem, and therefore have a rightful place on the global biological stage.

But what is the function of *Panthera tigris*? Why did Nature/God create it?

A poet would say that tigers add beauty, drama, and mystery to our world, which is true. A scientist, however, would say that tigers help *maintain the health and balance* of the natural world. Since this is a science book, let us look more closely at the second proposition, for from it we will see how tigers profit both humanity and our planet.

HOW TIGERS BENEFIT THE WORLD

Tigers play a vital role in the world ecosystem.

To begin with, as an apex predator the tiger helps *regulate* the population of wild herbivores, such as deer, buffalo, and other ungulates. Why is this important? Without tigers the overabundance of ungulates would result in great damage to the environment through the overconsumption of vegetation. This in turn would negatively impact the delicate equilibrium of the food chain, from the top down to the microbial level. While regulating herd populations, tigers simultaneously help rid ungulate gene pools of weak, sick, and aged individuals.

Thus, tigers act as a *natural check* on various imbalances, aiding in improving genetic stock, enhancing biodiversity, and maintaining the overall well-being of the ecosystem.

One often overlooked tiger value is that it provides a home and food for parasites, both internal (known as *endoparasites*) and external (known as *ectoparasites*). Among the parasites it hosts are a myriad of worm and tick species, each which serves its own important purposes (such as acting as food for other creatures, controlling wildlife populations, and—for us—serving as an ecosystem health indicator). Additionally, tigers target and moderate the populations of animals that we do not like, such as wild boars, which destroy untold millions of dollars of crops across Asia each year.

Tigers help control the overpopulation of ungulates, like these spotted deer.

Another benefit of having tigers in our world: the habitats tigers flourish in are also *vital to humans*. Grasslands, for example, contribute both to pollination and to the stabilization of the weather; marshes act as giant water filters; mangroves provide homes for newborn fish (known as larval fish) as well as young fish (known as fingerlings); and forests consume carbon dioxide while producing oxygen.

If tigers disappear from the wild, the biomes and ecosystems they once inhabited will be at risk for *development*, which would negatively impact the lives of not only people living in tiger countries, but people around the world as well. For all living things are connected in a single giant chain, the Web of Life, in which each link affects the ones on each side of it.

For these reasons a tiger is considered a *keystone species*: a type of animal that affects its environment in a profound manner; a manner that is out of proportion with its numbers, and which, if this species were removed, would cause significant deleterious changes in the surrounding

Palmetto scrubland provides homes for many organisms. Saving the tiger saves them as well.

ecosystem, such as loss of habitat and biodiversity.

The tiger is also an *umbrella species*: a species of plant or animal—usually occupying a large area—that when protected, helps ensure that co-occurring species of flora and fauna in its ecosystem will potentially be protected as well.

Naturally, in preserving tigers we are also helping *preserve* both the land they live on and the other animals (many of them endangered as well) that share the tiger's home. These in turn help maintain the beauty, products, enchantment, and wonder of the natural world that we humans enjoy so much.

Finally, there are the seldom discussed benefits of *education* and *ecotourism*:

- As captive zoo animals, tigers function as *educational models* and as *wildlife ambassadors*, giving voice to such issues as habitat loss and problems and resolutions associated with other endangered animals.
- As a focus of ecotourism, the profits that come from those who are willing to pay to see tigers in the wild can be refunneled back into *tiger conservation*, which includes reversing the longstanding entrenched notion that poached, trophy-hunted, dead tigers are more economically valuable than living wild tigers. Additionally, some of the money from ecotourism can go to local agrarian peoples, helping to raise their standard of living and improve their quality of life.

Tiger preservation, in turn, helps preserve valuable oxygen-producing forests like this one.

In the final analysis tiger conservation benefits far more than just tigers themselves. It is a boon to every living thing they are associated with—including us.

4

HOME & FAMILY

DIFFERENCES BETWEEN MALES & FEMALES

THE TIGER IS A *PHYSICALLY* dimorphic animal; that is, the two genders are different in appearance. Just as in humans, in the tiger's case the average male is larger, heavier, and more robust than the average female, who is smaller, weighs less, and has a lighter more gracile skeleton than her male counterpart.

THE RELATIONSHIP BETWEEN MALE & FEMALE

Unlike the gregarious lion which lives in a pride of up to 25 individuals, in comparison the tiger is something of a hermit, for it normally dislikes the company of its own kind. This makes it a *solitary animal*, one that only rarely assembles with other tigers. When it does, a tiger group is known as a "*streak*" or an "*ambush.*" Males and females do associate, but generally only during the mating season, and then only briefly, irregularly sharing resting spots and meals.

The male tiger (rear) and the female tiger (front) are physically dimorphic.

Tigers and tigresses are solitary animals, and rarely associate, except briefly during the breeding season.

Males may come into contact with one another just prior to the start of the breeding season, when they begin to compete over reproductive rights to estrus females. The strongest male is usually the victor of these contests, after which he guards his mate from competitors until her short estrus cycle comes to an end and they go their separate ways. Though the male-female relationship lasts only a few days, while they are together it is largely *patriarchal*, with the more powerful male dominating the female.

Outside the mating season male tigers and female tigers inhabit *separate territories*. When establishing their home range, *males* tend to concentrate on areas with the most abundant prey, the highest number of females, and the lowest number of rival males. *Females*, on the other hand, lean toward areas most suitable for raising young; that is, land that is rich in natural resources, such as food, water, trees, and good hiding cover. This means that the male's hunting grounds are often much larger than the female's.

How big is a male tiger's territory? Depending on a number of factors, including longitude, the season, the region, the ecosystem, and even the weather, it may be as small as three square miles (particularly when located near or around human settlements) or as large as 500 square miles (for example, in spacious uninterrupted wilderness areas).

While the male tiger will frequently tolerate females in his territory, other males are usually strictly forbidden, and those who break this unwritten law will be met with violent resistance from the feline land owner. Not all male tigers establish a territory, however. On rare occasions a particularly easygoing male will simply roam freely, casually intermixing with other males he may encounter along the way.

Tigers police their personal territory daily and nightly, marking off the *boundaries* with ground scrapes, tree and bush rubbing, tree scratching, and urine spraying (on rocks, bushes, and trees).

TIGER REPRODUCTION

Tigers do not have a set breeding season, which means they can *mate year round*. Despite this, a female tiger is only reproductively receptive for about three to five days every one or two months, with mating activity peaking between the fall and the spring—particularly in zones with moderate climates.

Tigers are *polygamous*; that is, they choose a new partner each time they reproduce, and thus have multiple mates throughout their lives. The tigress alerts males to her readiness with special *mating calls* and attracts them with peculiarly *scented urine*, which she deposits around her territory.

On average females become *pregnant* at around three years of age, while males begin *siring offspring* at around five years of age (rarely earlier). Female tigers are slow reproducers: with a *gestation period* of about three months, they bear a litter only once every three years or so.

TIGER BABIES

Around *90 days* after mating, the tigress gives birth to between one and seven wooly house cat-sized cubs, though two to three is average. Tiger neonates are *altricial*, meaning that they come into the world in a helpless state: blind, deaf, and utterly defenseless, and

Mother tiger and cubs.

weighing an average of only two lbs, they would quickly perish without parental care. Extra "baby stripes" provide camouflage for young tigers, additional protection against predators. These supplemental infantile markings disappear with age.

A few days after birth, the babies' eyes and ears start to open and function; at about two weeks tiny milk teeth appear. As the cubs grow, they are dutifully and attentively *nurtured*, fed, nursed, groomed, pampered, and fearlessly protected by their affectionate mother, who maintains them in a concealed den (such as in tall grass or a cave) that she selects and prepares a few days prior to parturition. Tiger mother care includes *licking* her infants' fur, which keeps them clean, strengthens the mother-baby bond, and induces healthy blood flow in their new tiny bodies.

Like the females of many other mammal species, mother tigers will *transfer* their offspring to a new home if the original one is exposed or disturbed by predators or people. While she is off hunting and eating, her highly active cubs play, tussle, and mock-fight, preparation for life as an adult. Throughout the *infancy period* they grow and gain weight quickly.

A mother tiger carries one of her cubs away from an approaching snake.

By two or three months of age the cubs start eating *solid food*, and by four or five months their permanent teeth grow in and they are *weaned*. Around this time the mother begins bringing them prey she has killed, while simultaneously teaching them how to hunt. She starts them off with small animals, a safe means of preparing them for taking down big game later in life.

Generally tiger cubs remain with their mother until between two and three years of age, by which time they have developed the size, strength, and skills to *solo hunt*. Passing on knowledge to the next generation (as opposed to learning exclusively by instinct) is one of the hallmarks of being a member of the mammal family.

Now nearly fully grown, this, the newest generation of *Panthera tigris*, attains independence and strikes out on its own. The life cycle of the wild tiger begins anew.

THE TIGER FAMILY & SOCIAL STRUCTURE

Tiger cub.

As is true of most mammal species, the tiger family does not include a father. Thus the *tiger nuclear family* is made up of only the mother and her offspring, making it *matriarchal* rather than patriarchal in nature. Additionally, the tiger nuclear family itself is *temporary*, as the young depart when they are mature enough to survive on their own, leaving the mother to her natural singular existence.

Living a solitary mode of life also means that tigers do not have the same kind of social structure found in communal animals, like wolves, who live in closely-knit packs of (sometimes several generations of) relatives. In fact, because they are so *reclusive* and *rare*, we do not know very much about the social structure of tigers—what little of it there appears to be.

THE TIGER PARENT

Because he leaves soon after mating with the tigress, a *tiger father* plays no role in the upbringing of his young, a task that is left solely to the mother. Nonetheless, before he departs, the father may aggressively defend his pregnant female from male rivals, which indirectly aids in the survival of the as-of-yet unborn cubs he has sired.

This illustration of a tiger family that includes the father is more fanciful than factual.

As for the *mother tiger*, we have seen how dedicated she is to motherhood and to insuring the well-being and survival of her infants. This includes defending her cubs to the death, viciously fighting off any creature that dares come too close or which she thinks might pose even a slight threat to them.

Incidentally, dangers include male tigers who, on occasion, will attack and kill cubs sired by other males. Though brutal, tiger populations can gain an evolutionary advantage from infanticidal males: the loss of her cubs causes a biochemical change in the mother, which often suddenly forces her into estrus again. This allows mating with the new male to take place. The DNA from this confident, assertive, dominant male is then passed on to the next generation through his cubs, strengthening the tiger gene pool.

TIGER SHELTERS

When they are not on the hunt, tigers usually spend their time out on *open ground* lounging, napping, and grooming. At other times a tiger will take up brief residence in a well hidden *den*, a temporary protective lair like a hollow log or cave, or underneath a thick hedge. Here it may rest, nap, relax, groom itself, and sleep for up to 20 hours a day.

Sometimes a tiger den may be nothing more than a small patch of shrubbery, a shady spot under a tree, a clump of bushes, a pile of rocks, or a large tussock. In biology these types of impermanent natural hideouts are known collectively as *hiding cover*: a screen of foliage, vegetation, or stone that is dense enough to conceal at least 90 percent of a creature from a distance of 200'.

With their orangish black striped coats, grasslands provide ideal hiding cover for *Panthera tigris*.

Hiding cover, which is used primarily by *shy* or *reclusive animals*, provides a tiger adequate concealment from its enemies, as well as a safe haven for females to rear their young. A tiger shelter is also sometimes used as a *hunting blind* from which to stalk and ambush prey.

A tiger mother and her three cubs relaxing in the morning sun.

5

DIET & HUNTING

THE TIGER DIET

TIGERS ARE *OBLIGATE CARNIVORES*, MEANING that they must eat meat in order to survive. Although a tiger will consume various plants if meat is unavailable, eventually it would perish without an adequate supply of flesh foods.

A tiger's menu incorporates everything from grass, berries, insects, crabs, fish, and rodents, to snakes, lizards, birds, amphibians, and rabbits. A more detailed list would include: deer, elk, water buffalo, crocodiles, wild cattle, tapirs, rhinoceros, foxes, turtles, wild boar, zebras, gaur (Indian bison), moose, rats, wild dogs, porcupines, monkeys, peacocks, jackals, antelope, elephants (mainly the young), wolves, frogs, ostriches, bears, leopards, and on occasion even its own kind—for cannibalism is not unheard of among tigers.

While the tiger's preferred food is *wild ungulates*, it will also take domestic livestock and human pets when they are available—particularly pet dogs. Other farm animals on the tiger menu include horses, cows, and

This large wild ox, known as a gaur, is one of the tiger's favorite foods.

goats, creating a problem that often spawns "revenge killings," in which villagers kill tigers in order to stop the killing of their animals. When live food is not available, tigers will feed on carrion, another benefit of tigers, as this helps keep the environment clean.

TIGERS AS MAN-EATERS

Debate continues as to whether we are part of a tiger's natural diet. On the one hand, our kind has shared the tiger's Asian domain for hundreds of thousands of years, plenty of time for it to have acquired a taste for human flesh. On the other hand, tigers seem to prefer nonhuman meat and rarely actually completely consume the many people they kill each year. This means that the term *"man-eater"* is not entirely appropriate, even when applied to those terrifying *rogue tigers* who occasionally prey on hundreds of people in quick succession.

Panthera tigris has a well-earned and enduring reputation as a man-eater.

We are gaining further insight into what may lay behind man-eating. For example, some evidence suggests that once a tiger develops a taste for domestic animals it may begin to see their human owners as food as well. Also, many tigers become man-eaters only because they are *aged, diseased, or injured*, or because they have *worn, broken, or missing teeth*. An animal with these problems will inevitably find catching and killing humans much easier than wild prey.

Overall, tigers prefer *solitude* and usually try to avoid humans. Indeed, the reality is that many if not most tiger-on-human attacks result from *diminishing tiger habitat*: as we encroach on their land it shrinks, becoming overcrowded with tigers; they then begin to venture outside the boundaries, bringing them into contact with people. Human interactions with problem tigers or "conflict tigers," as they are called, are always dangerous and often tragic.

While many mysteries still surround this phenomenon, it must be acknowledged that, as the great hunter and conservationist Jim Corbett once noted, the reasons most tigers become man-eaters are usually due to Man himself.

TIGER HUNTING HABITS

A tiger's favorite food, the wild ungulate, is mostly *nocturnal*. Thus the tiger is as well, with night being its preferred time to seek out its quarry. During the day, the tiger secludes itself in the shade of the forest to await the setting sun, at which time it emerges, lurking in wait along grassy paths, forest-bordered rivers, and the muddy banks of ponds and lakes. Despite its propensity for nightlife, tigers will hunt in broad *daylight* if need be, traveling up to 20 miles in a 24 hour period in search of a meal.

Unlike group-hunting carnivores (such as lions and wolves), a tiger is a *solitary ambush predator*: it hides in and moves through thick vegetation alone, usually tracking its victim from behind (but also sometimes from the side, above, or below), slinking as close to the ground as possible. It is always careful to remain absolutely *quiet* and *hidden* (behind rocks, bushes, trees, etc.). To increase their chances of success, in winter tigers prefer hunting in relatively snow-free areas, such as frozen ground patches, shorelines, and well used trails created by other animals.

Tiger taking down a deer.

When it is near enough to strike (normally within 25' to 75'), the tiger stealthily closes in then puts on a short burst of tremendous speed. This sprinting motion carries it forward to the point where it can leap onto its quarry. The shocking force of hundreds of pounds of feline muscle mass hitting the prey disorients it momentarily, giving the tiger just enough time to grab it by its *throat* or *neck*. It then delivers a fatal bite, breaking the animal's wind pipe or its neck bones.

To prevent being injured by its struggling prey (not an uncommon occurrence), it holds on tightly until the animal has *suffocated*. When all movement ceases and it is obvious that the prey has died, it is safe for the tiger to begin feasting. Tigers tend to begin eating from the rump area, moving forward toward the head region as they feed.

If the carcass is too big to be eaten at one sitting, the tiger hauls it to a secluded spot, *caching* the remains beneath vegetation (such as forest litter, leaves, branches, or grass). It will return over the next few days to consume the uneaten portions (even if they have begun to putrefy) until there is little left but the rejected parts (hair, hooves, etc.) and the skeleton.

It is important to note that, contrary to common belief, tigers rarely capture their intended prey. In fact, they are only successful around 5 percent of the time, giving it a 95 percent failure rate. If a tiger neglects to catch its victim after pursuing it for 100' or so, it will usually give up the chase. Its fortunate prey lives to see another day.

TIGER EATING HABITS

The shearing, cutting, and ripping functions of their razor-sharp teeth reveal that tigers do not chew their food as we—with our small, mostly flat-edged teeth—do. Instead, tigers tear off *large chunks* and swallow them whole.

A tiger can gulp down *70 lbs of meat* at one meal, the average weight of a bushel of corn, a Labrador Retriever, or a ten year old child. Very heavy eaters can pack away as much as 90 lbs at one sitting, the weight of a new born calf (cow), a large German Shepherd, or a 12 year old child.

This tiger waited patiently at a watering hole, then pounced on an Asiatic water buffalo that came to drink.

Its large muscular body, along with the extraordinary amount of energy it expends searching for food, means that a tiger must regularly consume large amounts of meat to stay fit and healthy. While it can live up to three weeks without food, it prefers eating every few days, and does not like to go more than a week without a large meal—which, on average, generally consists of about a 100 lb animal. Studies have shown that a tiger can eat about 50 such medium-sized animals a year, or approximately one per week. Thus a tiger needs and consumes around 5,000 lbs of meat a year.

THE TIGER'S SIX SECRET WEAPONS

When it comes to survival, I have identified what I call "the tiger's six secret weapons." Examining these helps us get closer to fully understanding the diet and hunting strategies of *Panthera tigris*. We will begin with number six and work our way upward to number one.

6. Weapon Number Six, Swimming: Tigers are one of the few members of the feline family that are attracted to water and who truly enjoy swimming. Thus, prey that flees into a body of water hoping to escape is almost certainly doomed, for, with their broad front paws and powerful forearms, tigers are *excellent swimmers* who can easily cross ponds, waterholes, rivers, estuaries, and even lakes as wide as 20 miles.

Tigers are superb swimmers, making them dangerous on land and in water.

5. Weapon Number Five, Generalist Eating Habits: Unlike specialized eaters, such as koala bears (whose narrow diet is restricted primarily to eucalyptus leaves), tigers are both *opportunistic and generalist eaters*. In other words, they will consume nearly any living thing that is edible and which they have an opportunity to kill. This type of diet permits tigers to take advantage of whatever prey is available, despite the terrain, season, weather conditions, or altitude.

Tigers will eat nearly anything they can catch, including crocodiles.

4. Weapon Number Four, Teeth and Bite Force: We have already discussed the tiger's impressive dentition, and how its 30 teeth can perform a variety of functions, such as cutting, slashing, gripping, snipping, and puncturing, as well as more delicate tasks such as plucking feathers and stripping meat from bones.

These carnivorous adaptations are further enhanced by the tiger's *bite force*, which, as we have seen, is one of the most powerful in the animal kingdom: at 1,050 psi, it can easily bite through tough animal hides, shred leathery sinew, and crack open thick bones. Its bite force is so strong that—combined with its heavily built dental weaponry—few animals are able to break its grip. Indeed, a tiger can drag a carcass twice its own body weight up to a mile. Only one big cat has a stronger bite: the jaguar can clamp down its jaws with a force of 1,350 psi.

Skull of a Bengal tiger, showing its crushing jaws.

3. Weapon Number Three, Agility: Despite their massive size and weight, tigers are *extremely nimble* creatures, with supple bodies that can twist and bend in nearly any direction. This allows them to perform acrobatic-like maneuvers while stalking, hunting, and killing, morphing and changing shape with the movement of their struggling prey. This snake-like ability greatly increases a tiger's chances for a successful meal, and is, no doubt, one of the reasons it long ago developed a legendary reputation as a *shape-shifter*.

2. Weapon Number Two, Speed: When a tiger launches an attack on its prey it can reach blistering velocities in only a second or two. Indeed, some subspecies can attain speeds of up to *40 mph* for short distances. This is fast enough to easily catch all but the most fleet-footed animals.

Extra powerful endurance muscles make the tiger both fast and strong.

1. Weapon Number One, Power: A tiger's gargantuan size (up to 14' in total length), its massive weight (up to nearly ½ ton in one subspecies), its imposing musculature (its front paws are capable of smashing a cape buffalo's skull), its prodigious set of specialized teeth (set like rows of steel daggers), and its exceptional bite force (nearly twice that of a lion), combine to create an imposing animal that, as a hunter, is nearly invincible.

6

COMMUNICATION

HOW TIGERS TALK TO ONE ANOTHER

LIKE MOST LAND MAMMALS, TIGERS communicate with conspecifics in a variety of ways that usually surround *sight* (visual signs), *touch* (tactile responses), *smell* (chemical odors), and *sound*. As we have discussed the first three modes in preceding chapters, this chapter will be devoted primarily to acoustical communication.

BASIC TIGER SOUNDS

Tigers are capable of making *thousands* of different *vocalizations* and are one of the world's most loquacious animals. Such sounds include: grunting, growling, moaning, hissing, snarling, mewing, purring, gasping, yelping, chuffing (a benign huffing sound also known as prusten), and, of course, roaring—the latter which, thanks to special skin flaps in the neck and throat, can be heard up to 5 miles away.

Panthera tigris can produce so many different sounds that we have yet to identify and catalog them all.

OTHER MEANS OF COMMUNICATION

They use other means of intraspecific communication as well, such as tail movements: a lowered limp tail, for instance, means a tiger is calm; a twitching tail indicates nervousness; a rapidly moving tail signals aggressiveness. As noted earlier, tigers also scrape the ground, spray urine on vegetation, and scratch trees in order mark off the boundaries of their territory, all overt threat signals to would-be intruders.

ULTRASONIC & INFRASONIC HEARING

One of the more fascinating aspects of tiger communication is linked to its ability to hear *ultrasonic* or *supersonic frequencies* (high) and *infrasonic* or *subsonic frequencies* (low), sounds well beyond what we humans can hear. Since most prey animals make or emit high-pitched sounds when they move (for example, leaf crunching and grass rustling, as well as vocal noises such as squeaking), being able to hear ultrasonic frequencies gives tigers an advantage when on the hunt.

Sensitive audio equipment, like this oscilloscope, has been used to determine the tremendous auditory capabilities of tigers.

COMMUNICATING & HUNTING SUBSONICALLY

A tiger's ability to hear in the infrasonic or low range is also necessary, and for a number of reasons. Tigers themselves produce *infrasonic sound* in order to communicate with conspecifics (its own kind). Though we cannot hear it, if we could it would sound like a low, deep, resonant rumble. In fact, we can physically *feel* subsonic sounds, which we sense as powerful vibrations that penetrate our bodies, seemingly vibrating our organs and bones. For what purposes does a tiger use low frequencies?

First, infrasound is used to *communicate* with other tigers, for example, as a friendly greeting, as a sign of reproductive readiness, as a threat to rivals, and as a way to beckon wayward cubs. Infrasound may also be used to locate other tigers in thick woodlands or over long

Tigers can hear and produce sounds far above and below human hearing.

distances, since low frequencies can pass through nearly any object, including trees, rocks, cliffs, gorges, hills, and even mountains.

Tigers use vocalizations and sound for a wide variety of purposes, from hunting to communication with conspecifics.

Second, low subsonic sounds are used when hunting to *surprise, distract, confuse, stun, and even paralyze* a tiger's prey. How? As a tiger lurches forward to attack, it emits a loud blast of noise. We call this terrifying sound a "roar," the lower sound spectrum which is composed of infrasound.

While most animals probably cannot hear the subsonic frequencies in a tiger's roar, they can *feel* the physiological effects, which they may experience as fear, panic, dread, ringing ears, nausea, or immobility. Hunting tigers take advantage of these painful and confused reactions in their quarry to move in for a quick kill.

This female tiger used infrasound to confuse her prey, then ambushed it from the side.

Sometimes the hunter becomes the hunted.

7

FRIENDS & FOES

THE ANIMAL ENEMIES OF TIGERS

AS APEX PREDATORS, TIGERS HAVE few natural enemies outside humans. One of these is the *dhole*, a sandy-reddish colored, undomesticated canine also variously known as the red dog, wild Asiatic dog, or Siberian wild dog.

Dholes, which normally prey on animals like lizards, rabbits, and deer, do not consider *Panthera tigris* prey. However, they do see it as a competitor for food and territory. And so, although the tiger outweighs an individual red dog by nearly a half ton, as a pack they will attack it and try to drive it off, and even kill it if possible.

The dhole.

The tiger has other foes, such as *snakes* and *leopards* (which may prey on young tigers), and the much larger *elephant* and the *Cape buffalo*. Though the latter two are herbivores and thus do not consume tiger meat, these enormous powerful mammals can seriously injure and kill tigers if they feel threatened.

Another major enemy of the tiger is *other tigers*; and more specifically, adult male tigers, who may prey on tiger cubs that they have not sired. As noted earlier, males do this in order to bring their mothers into estrus.

TIGERS AS AN ENEMY OF HUMANS

Tigers are one of the largest, fastest, strongest, and most cunning predators in the world. Thus when tigers and humans (and their livestock) cross paths, as they often do, it inevitably turns out badly for one or the other, or more typically for both.

The current policy for dealing with man-eaters, and also with tigers that show no fear of humans, is to kill them. How successful this policy has been for both tigers and humans is still being deliberated. However, its main purpose seems to be the temporary appeasement of local communities—and in this regard it has been somewhat successful. But this is not a viable long-term solution.

Humans and tigers have a long and troubled history.

Panthera tigris and *Homo sapiens* have shared the same habitats for many thousands of years, putting us directly in the tiger's sights and squarely on its menu. Indeed, tigers take the lives of more people every year than any other feline species, or any other mammal for that matter. They also kill and consume, in great numbers, farm animals, which, being largely rural, are invaluable assets to the people of Asia.

For these reasons tigers must be considered *one of our greatest wild animal enemies*: it is a true man-eater, one that has terrorized humanity for millennia and destroyed the lives, families, farms, and businesses of countless hundreds of thousands of people.

However, we must be careful not to demonize the tiger, especially conflict tigers. They are merely trying to live their lives as naturally as possible in what little land we have grudgingly given them. Remember, we are intruders in their home: we modern humans have been on earth for a mere 150,000 years or so, while tigers have been here for at least 2 million years.

HUMANS AS AN ENEMY OF TIGERS

As I will discuss in more detail shortly, the primary manner in which we hurt tigers is by the *ever-increasing growth of our population*, for to sustain more and more people each year requires the development of thousands of acres of new land.

This means building new towns and cities, along with the *infrastructure* needed to run them: roads, railways, bridges, tunnels, schools, stores, banks, hospitals, churches, farms, ranches, food production facilities, police and fire stations, water supply systems, telecommunication networks, sewage and waste disposal systems, airports, transportation systems, electric systems, power stations, landscaping, lighting, and dams.

Our kind is doubling in population every 50 years, requiring large swaths of new land for the building of cities. Much of this land is already inhabited by tigers, who are being driven into smaller and smaller wilderness areas.

Since tigers live largely on *undeveloped lands*, these are the regions developers usually buy up first, resulting in both tiger habitat loss and tiger habitat fragmentation. In the process, what becomes useable land to us becomes a deforested wasteland to tigers.

This new village road in India (lower right) cuts directly through the heart of tiger country.

One new road alone, carved through the middle of tiger country, is enough to disrupt the lives of all the tigers living in the area. How? By interfering with hunting, reproduction, and territorial boundaries, while limiting genetic variability (important to keeping a species healthy).

We have not even touched on human-made *water, air, noise, and light pollution*, and we need not, for the impact these forces have on tigers is obvious: just as with the expanding human population, they are nearly always negative.

WE ARE IN CHARGE OF THE TIGER'S FUTURE

This young tiger will need our help if his descendants are to survive into the 22nd Century.

Of course, just as we cannot demonize tigers for being tigers, we cannot demonize ourselves for being human. We will never stop or even slow the growth of human population. Thus, for better or worse, the destiny of the world's large carnivores is in our hands.

We must plan out the *tiger's future* in detail, for if we continue to follow the current trend, one day wild tigers will be gone, and their captive descendants will be relegated to parks, zoos, and nature preserves.

As we will see shortly, many individuals and organizations are already working diligently to guide and control the fate of *Panthera tigris* before it is too late. We will go from being the enemy of the tiger to becoming its friend, benefactor, and even savior.

This poached tiger skin was recovered by wildlife authorities in Nepal.

8

TIGERS IN SUPERSTITION, RELIGION, & LITERATURE

THE TIGER IN SUPERSTITION

BEING ONE OF THE MOST beautiful, charismatic, and awe-inspiring animals in the world naturally lends itself to becoming the focal point of superstition, which is precisely why *Panthera tigris* has generated so many folk beliefs over the past 5,000 years.

The tiger plays a central role in Asian culture.

Since the tiger is an Asian animal, the earliest folklore surrounding it is, naturally, Asian in origin. One of the more common folkloric beliefs is that revering the tiger brings *good luck*, particularly in regards to gambling. Early Chinese held the tiger to be a guardian spirit that *protected hunters and farmers*. A tiger's skin is considered quite sacred and is said to give an individual "a special power." Thus it is often used as a sitting mat for priests and other important officials, who make oaths and pledges while seated on it.

A symbol of fear and wonder due to its power, when a tiger roars, many infer that this forecasts the coming of storms and winds in the near future. Some maintain that the *tiger's roar* itself is the cause of "angry" weather. In many places it is still considered an ill omen if a tiger enters one's city, while wearing a moustache is said to make one "tiger proof." If a person is killed by a tiger, his or her family members will also become targets, so it is thought. To prevent another tragedy, the victim's kin must change their names to prevent the man-killer from finding them.

It is a taboo in many countries to kill a tiger, for there is a widespread belief that the *souls of one's ancestors* transfer into tiger bodies after death. Gardens are said to get their *fecundity* from wild tigers that live in the area. Killing a tiger then is sure to rain destruction down on people's vegetable plots and flower gardens.

In many parts of Asia it is believed that deceased ancestors occupy the bodies of tigers, making it a taboo to harm or kill them.

A tiger's whiskers are said to be toxic (and so are often used in poisons), while carrying a tiger's ear is thought to ward off sickness, wickedness, and even death. An image of a tiger placed on a doorjamb scares away *evil spirits*. In China, white tigers are not only associated with the merits of the aristocracy, but also with the feminine principle known as *yin*; hence, a troublesome woman is sometimes referred to as a "white tiger." Asians also link the tiger to the masculine principle, *yang*, due to its energy and vitality.

Just as the Irish avoid saying the word "fairies" for fear of offending them (instead they refer to them by such terms as "wee folk," "the gentry," or "good people"), in some regions of Asia uttering the word "tiger" is thought to bring *bad luck*. In China, for example, the word tiger, spelled *hu*, is replaced by various nonsense phrases, such as "the giant reptile," "the Lord of the Black Rock," or "the king of the mountain."

Tiger meat is thought to have a *medicinal effect* on the body, while tiger fangs and claws, when made into potions, can attract love and fortune. *Respect* for the tiger is believed to procure its protection, thus in some areas people consider it bad luck to threaten tigers by pointing guns or even spears at them. Instead they are honored and revered.

TIGERS IN RELIGION & SYMBOLISM

Tigers have been among us for a long time, which is why they appear on the *stone tablets* and *seals*, and in the *rock and cave art*, of societies living as far back as 5,000 years ago.

It was *Alexander the Great*, who lived during the 4th Century BC, who seems to have been the first Westerner to observe and record a wild tiger, a memorable event that took place during his campaign to India. Some 1,300 years later, during the 12th Century, *King Henry I* became the probable first owner of a tiger in England, a curiosity which aroused great interest in Medieval Europe.

Tigers were frequent guests in ancient Rome, where they were used for sport and entertainment.

In ancient times the *Roman Empire* employed captured tigers for sport, entertainment, exhibitions, and executions. Around 20 years before the birth of Jesus, Indian tourists bestowed a tiger on the Roman *Emperor Augustus*, who must have noticed the *religious zeal* with which tigers were worshiped by Easterners. Indeed, this tradition can still be found across nearly all of Asia, where they are celebrated and worshiped as, for example, the tiger deities Dakshin Ray and Baghdeo. There is good reason for this.

Like most lifeforms and objects, the tiger represents both positive and negative qualities, on the one hand *power* and *stamina* and, on the other, *mercilessness* and *ferocity*. Thus it has a long history of dualistic religious symbolism attached to it. For example, while it is sometimes used to portray *evil or darkness*, its seemingly magical ability to find its way through shadowy foreboding woods also makes the tiger a symbol of the *Christ* or inner *light* (see John 1:9 KJV), which helps lead humanity through the murky twilight of gross materiality.

Western religion has found many other uses for the tiger, such as its function as a symbol related to *divinities*. Due to their majestic personas, for instance, the ancient gods and goddesses were said to ride on tigers as if they were horses. The tiger's wild fury causes even the Devil to quake in fear, and so it is often carved into *headstones* to frighten away demons. Naturally, artistic depictions of tigers have graced the crowns, flags, and coins of countless *nations* around the world since time immemorial, while early warlords liked to appropriate the tiger's image as an emblem of their courageous leadership.

The *chief deities* of ancient Mediterranean peoples are depicted not only sitting on tiger skins, but also wearing them, as they are numinous symbols of authority, potency, resilience, aristocracy, audacity, nobility, and bravery. The tiger has an age-old connection to specific divinities, such as the Phrygian mother goddess Cybele, the Greek wine god Dionysus, and the Latin wind god Favonius.

The ancient Roman mother-goddess Cybele was associated with numerous animals, including the genus *Panthera*.

In the northeastern Indian state of Assam, it is said that Budhi Pallien the forest-goddess manifests in the shape of a tiger, while China's most important ancient goddess, Hsi Wang Mu, was imaged as a monstrous deity with a wild mane and tiger's teeth. The tiger is also revered as a symbol of *healing*, for many maintain that an emblem of a tiger can absorb illnesses, especially in children.

In the *Chinese calendar* every twelfth year is recognized as "the Year of the Tiger," while in *Chinese astrology* the third sign in the Zodiac is named after the tiger (the equivalent of the Western zodiacal sign Gemini).

The people of Vietnam, Laos, Malaysia, Cambodia, Indonesia, and Korea all idolize the tiger in a variety of forms and for a myriad of reasons. For instance, it is a *protector* against nightmares, danger, and evil, as well as a supernatural being with many *magical powers*—such as the ability to shape-shift. *Weretigers* (like werewolves) are said to be able to transform themselves from animal to human and back again (a mysterious talent known as therianthropy).

China's *four-quarter astrological signs*, which correspond to the four cardinal directions, are made up of a black tortoise (north), a red bird (south), a blue dragon (east), and a white tiger (west). Early Chinese people held that the wind is created by the breathing of a tiger. In the Book of Ezekiel, *Judaism's* four-quarter astrological signs are called the "four living creatures," which, in Christianity, were later associated with the *Four Evangelists* (a symbol known as the Christian "Tetramorph"), and even later, the Four Archangels of the Book of Revelation (7:1).

As a symbol of ungovernable passions and emotions, in early *Christianity* the tiger was an emblem of both a cunning animal and an evil beast, thereby linking it with Satan. Like Lucifer, a tiger could trick the mentally weak and the spiritually indolent by posing as a beautiful and enticing creature. When its victim drew close enough, the Devil-as-tiger would pounce on it and tear it to pieces.

One ancient belief held that *tiger hunters* could gain an edge on their dangerous quarry by appealing to a mother tiger's well-known powerful maternal instinct: after stealing away one of her cubs, she angrily gives chase. But the crafty human hunter leaves a mirror on the road behind him. Coming upon it the tigress peers into it. Mistakenly thinking her own reflection is her cub, she prepares to nurse it. As she lays down, her human pursuer leaps from the bushes and unloads his weapon.

TIGERS IN POPULAR CULTURE & SCIENCE

With its many striking attributes, the tiger is a common figure in *popular culture*, appearing, for instance, on cereal boxes ("Tony the Tiger"), in songs (*I've Got a Tiger by the Tail*), and in movies ("Rajah" in *Aladdin*).

The tiger has been an eponymous contributor to *American sports*, lending its name, for instance, to such teams as the Tennessee State Tigers, the West Alabama Tigers, the Detroit Tigers, the Clemson Tigers, the Texas Southern Tigers, the Cincinnati Bengals, the Memphis Tigers,

Lilium tigrinum: the tiger lily.

the Princeton Tigers, and the LSU Tigers.

The tiger is also well represented in various *scientific fields*, such as botany (the tiger lily), entomology (the tiger moth), geology (the tiger's eye, a gemstone), ichthyology (the tiger shark), herpetology (the tiger salamander), and ornithology (the tiger swallowtail), to name but a few examples.

TIGERS IN LITERATURE

Nowhere does the tiger appear more often than in literature, from novels and poetry to children's books and real life adventures.

English author A. A. Milne included one named "Tigger" in his famous Winnie the Pooh series, while English novelist Rudyard Kipling created a fictional tiger called "Shere Khan" for his celebrated children's book *Jungle Stories*. In his poem *Gerontion*, American-English author T. S. Eliot makes reference to "Christ the tiger," while British writer-hunter Jim Corbett sensationalized *Panthera tigris* in his riveting real life jungle book entitled, *Man-eaters of Kumaon*.

William Blake authored one of the world's best known tiger poems.

In his famous dramatic play *Henry V*, William Shakespeare placed the tiger in a military context:

> But when the blast of war blows in our ears,
> Then imitate the action of the tiger;
> Stiffen the sinews, summon up the blood,
> Disguise fair nature with hard-favour'd rage:
> Then lend the eye a terrible aspect . . .

While hundreds of examples of such literary tigers could be cited, arguably the most famous Western reference to the tiger is a poem by the English Christian mystic William Blake. Written in 1794, it is entitled simply, *The Tyger*:

> Tyger! Tyger! burning bright

In the forests of the night,
What immortal hand or eye
Could frame thy fearful symmetry?

In what distant deeps or skies
Burnt the fire of thine eyes?
On what wings dare he aspire?
What the hand dare seize the fire?

And what shoulder, and what art,
Could twist the sinews of thy heart?
And when thy heart began to beat,
What dread hand? and what dread feet?

What the hammer? what the chain?
In what furnace was thy brain?
What the anvil? what dread grasp
Dare its deadly terrors clasp?

When the stars threw down their spears
And water'd heaven with their tears,
Did he smile his work to see?
Did he who made the Lamb make thee?

Tyger! Tyger! burning bright
In the forests of the night,
What immortal hand or eye
Dare frame thy fearful symmetry?

TIGER SAYINGS, EXPRESSIONS, & PROVERBS
With their many spectacular traits, tigers have also long been the subject of proverbs and sayings. Some samples:

"He who rides the tiger can never dismount." — Chinese proverb

"Even when a girl is as shy as a mouse, you still have to beware of the tiger within." — Chinese proverb

"It is not part of a true culture to tame tigers, any more than it is to make sheep ferocious." — Henry David Thoreau

"Don't strike a flea on a tiger's head." — Chinese proverb

"Being a president is like riding a tiger. You have to keep on riding or be swallowed." — Harry Truman

"The tiger crouches before he leaps upon his prey." —Timothy Pickering

"To beat a tiger one must have a brother's help." — Chinese proverb

"One day as a tiger is worth a thousand as a sheep." — Chinese proverb

"Once Confucius was walking on the mountains and he came across a woman weeping by a grave. He asked the woman what her sorrow was, and she replied, 'We are a family of hunters. My father was eaten by a tiger. My husband was bitten by a tiger and died. And now my only son!' 'Why don't you move down and live in the valley? Why do you continue to live up here?' asked Confucius. And the woman replied, 'But sir, there are no tax collectors here!' Confucius added to his disciples, 'You see, a bad government is more to be feared than tigers.'" — Lin Yutang

A tiger can drag a carcass twice its own body weight up to one mile.

"Unless you enter the tiger's den, you cannot take the cubs." — Asian proverb

"When a man wants to murder a tiger he calls it sport; when a tiger wants to murder him, he calls it ferocity." — George Bernard Shaw

"It is no use trying to satisfy a tiger by feeding him with cat's meat." — Winston Churchill

A tiger's front footprint or pugmark. Like a human fingerprint, every tiger pugmark is unique.

"A tiger cannot beat a crowd of monkeys." — Chinese proverb

"A tiger dies and leaves his skin; a man dies and leaves his name." — Japanese proverb

"The most difficult thing is the decision to act, the rest is merely tenacity. The fears are paper tigers. You can do anything you decide to do. You can act to change and control your life, and the procedure, the process is its own reward." — Amelia Earhart

"Tiger and deer do not walk together." — Chinese proverb

"Do not blame God for having created the tiger, but thank him for not having given it wings." — Indian proverb

From Hinduism to Christianity, tigers have always been popular figures in religious myth, ritual, and symbolism.

Captive tigers, like these two in Italy, are permanent citizens at zoos, parks, and reserves around the world.

9

VIEWING & IDENTIFYING TIGERS

IN SEARCH OF THE WILD TIGER

THE VAST MAJORITY OF US will never have the amazing experience of witnessing a *wild* tiger in its natural habitat. The only tigers most people glimpse are on TV, or, if they are fortunate, a *captive* tiger at a zoo.

There are two obvious reasons for this: wild tigers only live in Eurasia, and wild tigers are extremely scarce.

But what if you have the means and the time to go to a location where wild tigers live? Or what if you live in Eurasia? Then this next section is for you. Let us begin this chapter with a look at a few of the more famous places for observing *wild tigers*.

Seeing a tiger in person, particularly a wild tiger, is always a thrilling experience.

WHERE TO SEE WILD TIGERS: NATIONAL PARKS & RESERVES

Though rare, wild tigers range over a wide, extremely rugged, and often inaccessible area of Asia and eastern Europe, one covering many hundreds of thousands of square miles. This makes seeing one in the flesh a rather difficult task.

The following partial list of parks, however, will make this goal a bit more attainable, as they cater to *wild tiger* tourists and enthusiasts who either live in Eurasia or are willing to travel to Eurasia:

- Chitwan National Park: Located in Chitwan, in the subtropical region of Nepal, this World Heritage Site protects some 70 mammal species, one of which is the royal Bengal tiger. Seven resorts inside the park accommodate guests, who may take an elephant safari in the hopes of spotting a wild tiger.
- Zov Tigra National Park: This 200,000 acre refuge, located in Primorsky Krai (4,000 miles east of Moscow), is Russia's first national park set aside specifically for tigers. The park, whose name is Russian for "roar of the tiger," is known for its ecologically-based nature tourism, which gives visitors a chance to see a tiger in the wild.
- Sunderbans Tiger Reserve: Possessing some 350 wild tigers within its secluded borders, this forested mangrove park, located in West Bengal, India (on the border with Bangladesh), has the distinction of hosting one of the largest populations of *Panthera tigris* in the world. An added feature are its boat safaris, which comb the park's rivers in search of swimming tigers.

Swimming tiger, Sunderbans Tiger Reserve.

- Ranthambore National Park: Known as "the Home of Royal Bengal Tigers," tigriphiles, birdwatchers, and wildlife photographers flock to this popular destination, located in Rajasthan, India (about 250 miles from the city of Delhi), to enjoy guided tours from the safety of jeeps and trucks.

- Kerinci Seblat National Park: The largest national park on the island of Sumatra, Indonesia, the rain forests of this massive reserve contain several hundred Sumatran tigers. You might see one, along with gibbons, hornbills, and an assortment of other wildlife, on the park's famous 5-day "Sumatran Tiger Trek," which takes you deep into the beautiful Sumatran wilderness.
- Bandhavgarh National Park: Feline lovers visiting this spot have the opportunity of seeing not only tigers, but a myriad of other big cats, such as leopards. Located in Madhya Pradesh, India, the park boasts tiger safari tours that launch several times a day from Mahua Kothi, a 50-acre luxury resort, complete with gardens, a swimming pool, nature walks, birding trips, and more.

WHERE TO SEE CAPTIVE TIGERS: ZOOS & SANCTUARIES

What follows is a partial list of some of the more popular zoos, parks, preserves, and sanctuaries that house *captive tigers*. Many of these are generic tigers as opposed to the pure tiger stock found in the wild. Nonetheless, seeing a living tiger, whatever its genetic background, is *always* an awe-inspiring experience:

Hundreds of zoos and sanctuaries maintain captive tigers, making these rare big cats widely available for public viewing.

WITHIN THE U.S.A.
- Cincinnati Zoo: Cincinnati, Ohio.
- Memphis Zoo: Memphis, Tennessee.
- Big Cat Rescue: Tampa, Florida.
- Nashville Zoo: Nashville, Tennessee.
- Philadelphia Zoo: Philadelphia, Pennsylvania.
- San Diego Zoo Safari Park: San Diego, California.
- Dade City's Wild Things: Dade City, Florida.
- Oakland Zoo: Oakland, California.
- Bronx Zoo: Bronx, New York.
- Roar Foundation/Shambala Preserve: Acton, California.
- Hogle Zoo: Salt Lake City, Utah.

- Sierra Nevada Zoological Park: Reno, Nevada.
- Brookfield Zoo: Brookfield, Illinois.
- Toledo Zoo and Aquarium: Toledo, Ohio.
- Detroit Zoo: Detroit, Michigan.
- Zoo Knoxville: Knoxville, Tennessee.
- Tiger World: Rockwell, North Carolina.
- Carolina Tiger Rescue: Pittsboro, North Carolina.
- The Wild Animal Sanctuary: Keenesburg, Colorado.
- Great Cats World Park: Cave Junction, Oregon.
- The Alaska Zoo: Anchorage, Alaska.
- Black Pine Animal Sanctuary: Albion, Indiana.
- Tiger Creek Animal Sanctuary: Tyler, Texas.
- Omaha's Henry Doorly Zoo and Aquarium: Omaha, Nebraska.
- Rio Grande Zoo: Albuquerque, New Mexico.
- Oklahoma City Zoo: Oklahoma City, Oklahoma.
- Denver Zoo: Denver, Colorado.

OUTSIDE THE U.S.A.
- Kristiansand Zoo and Amusement Park: Kristiansand, Norway.
- Copenhagen Zoo: Copenhagen, Denmark.
- Kolmården Wildlife Park: Kolmården, Sweden.
- Korkeasaari Zoo: Helsinki, Finland.
- Lujan Zoo: Lujan, Argentina.
- Zoo Pomerode: Pomerode, Brazil.
- Zoo Brazov: Brazov, Romania.
- Cabárceno Natural Park: Cantabria, Spain.
- Selwo Aventura: Estepona, Spain.
- Lo Zoo di Napoli: Naples, Italy.
- Attica Zoological Park: Spata, Greece.
- Australia Zoo: Beerwah, Australia.
- Taronga Zoo: Sydney, Australia.
- Jungle Cat World Wildlife Park: Clarington, Canada.
- Assiniboine Park Zoo: Winnipeg, Canada.
- Magnetic Hill Zoo: Moncton, Canada.
- London Zoo: London, England.
- Dudley Zoo and Castle: Dudley, West Midlands, England.
- Howletts Wild Animal Park: Littlebourne, England.

- Hamerton Zoo Park: Steeple Gidding, England.
- Isle of Wight Zoo: Yaverland, England.
- Welsh Mountain Zoo: Colwyn Bay, Wales.
- Dublin Zoo: Dublin, Ireland.
- Magdeburg Zoo: Magdeburg, Germany.
- Tierpark Berlin: Berlin, Germany.
- Berlin Zoo: Berlin, Germany.
- Eifel Zoo: Lünebach, Germany.
- NaturZoo Rheine: Rheine, Germany.
- Parc des Félins: Lumigny-Nesles-Ormeaux, France.
- Sofia Zoo: Sofia, Bulgaria.
- Novosibirsk Zoo: Novosibirsk, Russia.
- Leningrad Zoo: Leningrad, Russia.
- Izhevsk Zoo: Izhevsk, Russia.
- Moscow Zoo: Moscow, Russia.
- Rostov Zoo: Rostov on Don, Russia.
- Kaliningrad Zoo: Kaliningrad, Russia.
- Ueno Zoo: Tokyo, Japan.
- Shirotori Zoo: Matsubara, Japan.
- Zoorasia: Yokohama, Japan.
- Tobu Zoo: Miyashiro, Japan.
- Fuji Safari Park: Shizuoka, Japan.
- Johannesburg Zoo: Johannesburg, South Africa.
- Pretoria Zoo: Pretoria, South Africa.

Controversy surrounds the world of captive tigers. But zoos and sanctuaries are still the only way most people will ever be able to see a living *Panthera tigris*.

HOW TO IDENTIFY A TIGER

Whether in the wild or in a zoo, distinguishing a tiger from other animals, especially other felids, could not be easier: not only is it the largest cat in the world, it is the only big cat with its coloration and markings: its massive size, black stripes, and orangish coat make it completely unique in appearance among large felines. Details on the specific identifying characteristics of each of the 12 tiger subspecies are provided in Chapter Eleven.

THE TIGER-WATCHING KIT

Tiger-watching requires the appropriate tools, which means putting together a tiger-watching kit. And the type of kit you put together will depend, to a certain extent, on whether you are an amateur or a professional, as well as whether or not the tigers you will be watching are in an enclosure or in the wild. For simplicity's sake I will combine both kits into one.

A pair of binoculars should be part of every tiger-watcher's kit.

The number one tool everyone will need is a set of *binoculars*. These will let you view the details of, for example, a tiger's teeth, its ears, its eyes, and its markings, particulars that you would not otherwise be able to make out clearly with the naked eye. For getting in really close to your subject nothing is superior to a *telescope* mounted on a tripod, especially for tigers that are sitting or sleeping.

The next item on the list is a *recordkeeping device* of some kind, such as a pen and notepad, digital recorder, cell phone, tablet, or a notebook or laptop computer. Maintaining records of your sightings, as well as the date, time, and weather conditions of your trips (to the zoo or the wilderness), are vital for both amateur and professional tiger-watchers.

If you are lucky enough to come across a tiger in the wild, you will need to *document* where and when you saw it, along with both its common name and scientific name. Describe the cat in as much detail as possible, carefully recording its behavior, its markings, its approximate age, and whether it is a female or male. To make your record even more scientifically useful, chronicle the temperature, humidity, overall weather conditions, elevation, time of day, and exact location, as well.

A camera allows you to catalog and review your tiger photos at your leisure.

Do not forget to keep a *tiger log*, a running list of every tiger you see. You will want to keep track of how many different tiger subspecies you see each year, and indeed, throughout your entire life. In Chapter Eleven I have provided a small box in front of each living tiger's common name, where you can place a check mark when you have seen and identified that particular subspecies.

Choose a tiger-watching backpack that fits your individual needs.

Finally, you will need a still *camera* or a *video camera* to record your sightings. If you have a camera that allows for interchangeable lenses, a telephoto lens would be a useful addition, as well. Your photos will not only allow you to study your sightings more carefully later on, but it will also give others a chance to share in the excitement of your tiger experiences.

Store everything in a *backpack* for your trip to the zoo or wildlife sanctuary. For those venturing into the wilderness, include standard pieces of hiking equipment (flashlight, matches, insect repellant, compass, GPS tracker, water filter, personal locator beacon, etc.).

With your list completed you will be well prepared for any tiger you come across, free or captive!

Due to their physical and behavioral similarities, circuses often train tigers (right) and lions (left) together.

While provocative, this imaginative illustration portrays a felid event that has never occurred in nature, and never will: a fight over a deer kill between a wild tiger, a wild lion, and a wild jaguar. The reason? *Panthera tigris* lives in Eurasia, *Panthera leo* lives in Africa, and *Panthera onca* lives in the Americas.

10

CARING FOR OUR FELINE FRIENDS

ALL TIGERS ARE PROTECTED BY LAW

YEAR BY YEAR, DAY BY day, the six remaining living tiger subspecies are quickly disappearing from the wild, which is why steps to preserve them have long been underway. As tiger numbers continue to fall, new regulations have been created to help stem the trend toward extinction—including listing three of them as *endangered*, and the other three as *critically endangered*.

What follows are just a few of the global laws that help, directly and indirectly, protect both wild and captive tigers:

- The Animal Welfare Act (1966).
- Project Tiger (1973).
- The Endangered Species Act (1973).
- The Convention on International Trade in Endangered Species of Wild Fauna and Flora (1973).
- Captive-Bred Wildlife Regulations (1979).
- The Forest Conservation Act (1980).

- The People's Republic of China's Wildlife Animal Protection Law (1988).
- The Rhinoceros and Tiger Conservation Act (1994).
- The Captive Wildlife Safety Act (2003).
- The Big Cat Public Safety Act (2015-2016).
- The Wildlife Conservation and Anti-Trafficking Act (2019).

WHAT IS PUSHING WILD TIGERS TOWARD EXTINCTION?

The heaviest impact on *Panthera tigris* is the ever burgeoning population growth of *Homo sapiens*.

It has been estimated that in the past century alone, the worldwide tiger population has dropped at least 97 percent, making it *the most endangered species of wild cat on earth*. Some believe that at this rate all wild tigers will be gone in 30 years (2050). At least six tiger subspecies are already extinct, and the remaining six are traveling stubbornly down the same road.

There are a myriad of reasons for this. Arguably the most impactful one is *human population growth*, which, worldwide, is currently doubling every two generations (around every 50 years).

It is a simple matter of *space and resources*: as more people are born, more land and food is required to maintain them. Tigers, being large carnivores, need abundant sources of food and water, as well as large, undisturbed, contiguous wild spaces in which to live, reproduce, and hunt. Thus the number of humans has a direct and immediate effect on the number of wild tigers: the more people there are, the less land and resources there are for tigers. In short, *more people, less wild tigers*.

The city of Kolkata (Calcutta), India, current human population: 15 million—and growing.

There are many other threats to the survival of wild tigers as well:

- Habitat destruction: The grasslands and forests tigers prefer are disappearing at an alarming rate. This is due, in great part, to the *development* of housing, urban, commercial, industrial, recreational, and tourist areas. Farms, ranches, mining, warfare, dams, and human-made fires, also take their toll on tiger land. Another important source of habitat loss is the timber industry, which contributes to declining tiger numbers by destroying large swaths of their habitat. Indeed, *deforestation* is today one of the primary pressures on tigers, pushing them ever closer to complete eradication. Some estimate that at least 90 percent of tiger habitat has disappeared in just the last 20 years (that is, between the years 2000 and 2020).

Habitat destruction, to make way for new cities, means that tiger lands are receding rapidly.

- Trophy hunting: Though not as common as it once was, this "sport" helps deplete the tiger's already decreasing gene pool. It is not all bad news, however. *Ethical* hunting practices (a code of conduct for hunters which follows a strict set of moral and legal guidelines) are reducing the impact hunters are having on wild tigers.
- Tiger parts market: In many locations of the world various parts of the tiger, such as its skeleton, head, bones (mainly ribs), skin, jaws, claws, teeth, skull, paws, legs, reproductive organs, and tail, have been used in *folk medicine* for thousands of years. Seen mainly as curatives, they are believed to do everything from eliminate acne, toothaches, burns, dysentery, asthma, and malaria, to mental illness, tuberculosis, ulcers, rheumatism, typhoid fever, and epilepsy.

Hunter posing with a tiger kill.

Additionally, tiger skin rugs, tiger skin pillows, tiger skin cushions, tiger skin wall hangings, and winter wear made from tiger fur are still used in some regions, tiger skins and stuffed tigers are often employed as opulent home decor, its bones are turned into wine, its whiskers are ground into an alleged toxic powder, and its teeth and claws are used as talismans and jewelry. Such items are regarded as *status symbols* by their users and owners. Finally, *tiger meat* is in great demand by the wealthy as a luxury food. Much of this industry surrounds *illegal smuggling*. For many, the illicit tiger parts trade is a tempting one: sellers command high prices, sometimes making the equivalent of over a year's salary for a single pound of tiger bones—which can sell for up to $100. A complete tiger body can fetch nearly $100,000.

Tiger hunting has long been illegal worldwide. Yet the sport is on the rise, for as tigers become rarer their value increases.

- The Pet Trade: Thousands of tigers are kept as *exotic pets*—many illegally. Debate concerning the merits, demerits, and ethics of this always potentially deadly pastime rages on. For instance, some tigers live with their owners in small urban apartments. Is this animal abuse? Additionally there is the fact that a number of tiger owners have been seriously injured or even killed by their "pets" over the years. And what about the possibility of a city-dwelling escapee? A tiger roaming loose in a metropolitan area poses a serious danger, and would certainly end up being killed as a result. Should private tiger ownership be more tightly regulated, or even outlawed, due to its many inherent hazards?
- Poaching: Many tigers are lost to *illegal trapping and hunting*, wildlife crimes that are usually motivated by the following: fear of tigers (often very real), an effort to stop tigers from killing domestic livestock (known as "revenge killing"), a need for food (common in poor regions), the illicit tiger parts trade, the unlawful tiger breeding black market, and the demands of traditional medicine. Corrupt government officials, who may take money in exchange for granting access to restricted tiger lands, add to the problem.

- Declining food supply: As the human population expands, we take more and more *resources* from the environment instead of judiciously sharing them with tigers. The result may be unintentional, but it is unavoidable.

HOW YOU CAN HELP TIGERS

It is obvious that tigers need our assistance if they are to survive into the future, particularly since we are the primary reason their numbers are dwindling.

The logging industry takes a great toll on tiger habitat, but it is also a renewable resource that greatly benefits us—just one of the many dilemmas posed by tiger conservation.

But helping tigers is not as simple as helping, for example, owls, which are small, fairly common animals that are global in distribution. Tigers, in contrast, are large, rare, dangerous animals that live only in Eurasia. Thus, for most of us, any assistance we offer must be indirect; that is, channeled through the work of others.

Arguably the most important thing a *Panthera tigris* lover can do to aid tigers is to join and support legitimate, trustworthy, ethical, non-political (and non-politicized) conservation groups. Such organizations use the money you donate to fund their work with politicians, scientists, police, game wardens, local governments, village communities, and hunters and hunting organizations.

A dead conflict tiger.

Tiger preservation is an expensive, complex, labor- and time-intensive process that involves and requires:

- Thousands of people (both trained professionals and untrained amateurs).
- Office facilities.
- Population management and analysis.
- Resource management.
- Constant new legislation.

- Strengthening and reinforcement of existing regulations.
- Countless man- and woman-hours patrolling and guarding tiger lands, sanctuaries, and reserves from illegal hunting and trapping.
- Field research.
- Training forest guards and patrols.
- Equipment, such as radios, cameras, firearms, and vehicles.
- Regular inspections of facilities.
- Field data collection.
- Surveys of tiger populations and their prey.
- Habitat assessment and preservation.
- Improving zoo facilities.
- Protection of prey bases.
- Setting up emergency rapid response funds.
- Setting up conservation trust funds.
- Support networks for NGOs (non-governmental organizations).
- Computer analysis.
- Tiger census and monitoring programs.
- Anti-poaching operations.
- Radio-collars.
- Breeding strategies.
- Ongoing public education.
- Establishing tiger reserves, tiger development zones, tiger buffer zones, and tiger corridors.
- Wildlife studies.
- Workshops.
- Conservation activities.
- Community outreach programs.
- Hunting control.
- Enlarging nature reserves.
- Repopulation and resettlement.
- Eliminating demand for tiger products.
- Shutting down markets for tiger products.
- Promoting alternatives to tiger products.
- Conservation awareness projects.
- Managing tiger ecotourism.
- Recycling profits from tiger ecotourism back into tiger conservation.

Russian postage stamp.

Poachers skinning a tiger for the illegal tiger parts trade.

By partnering with *reputable conservation groups* (I say reputable because some conservation organizations are little more than fronts for extremist political groups), you help facilitate the ongoing efforts to save wild tigers, which, as noted, include saving tiger habitat, protection against poachers, strengthening anti-hunting laws, offering economic incentives to tiger-range towns, stronger policing of illegal tiger trafficking, urging cross-governmental cooperation, and the development of alternatives to tiger parts and their associated products—among many other important tiger-saving activities.

Highly trained tiger protection units, like this group being deployed in a Eurasian national park, require considerable funding.

Time is essential. If nothing is done, it is estimated that by the year 2050 every wild tiger subspecies will have vanished. Only an immediate and concerted global effort can save the six remaining subspecies of *Panthera tigris* from complete extermination in the wild.

This tiger has been tranquilized so that it can be fitted with a radio collar, allowing tiger researchers to study its habits in greater detail.

Tigers are territorial and will attack intruders much larger than themselves.

11

A TIGER-WATCHER'S GUIDE TO PANTHERA TIGRIS

HAVING STUDIED THE BASICS OF tiger science, it is time to take a closer look at the living tigers you are most likely to see, both captive and wild. It is also pertinent to familiarize ourselves with those tigers who have already lost their battle with humanity and Nature and are no longer with us.

The following is an inventory of all 12 subspecies, listed in alphabetical order according to their common names, and accompanied by color photos. Study it and familiarize yourself with each cat.

When you next visit a zoo that has tigers, or the next time you go out into the field to tigerwatch, take this book along and use this chapter as an identification guide. Put a check mark in the box in front of a tiger's common name once you have seen and identified it. This is your "life list," and you will find yourself referring to it often as your tiger-watching experience grows.

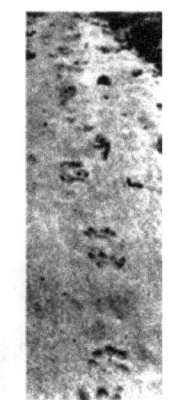
Tiger pugmarks.

ALPHABETICAL INDEX TO THE TIGERS IN THIS CHAPTER

12 Eurasian Tiger Subspecies: Six Living, Six Extinct

1. Amur tiger (*Panthera tigris altaica*) - living.
2. Bali tiger (*Panthera tigris balica*) - extinct.
3. Bengal (or Indian) tiger (*Panthera tigris tigris*) - living.
4. Caspian tiger (*Panthera tigris virgata*) - extinct.
5. Indochinese tiger (*Panthera tigris corbetti*) - living.
6. Javan tiger (*Panthera tigris sondaica*) - extinct.
7. Malayan tiger (*Panthera tigris jacksoni*) - living.
8. Ngandong tiger (*Panthera tigris soloensis*) - extinct.
9. South China tiger (*Panthera tigris amoyensis*) - living.
10. Sumatran tiger (*Panthera tigris sumatrae*) - living.
11. Trinil tiger (*Panthera tigris trinilensis*) - extinct.
12. Wanhsien tiger (*Panthera tigris acutidens*) - extinct.

Circus tiger in training.

Wild Amur tiger on the prowl.

AMUR TIGER

☐ 1. COMMON NAME: **AMUR TIGER**
ALTERNATE COMMON NAMES: Siberian tiger, Manchurian tiger, Ussurian tiger, Northeast China tiger.
SCIENTIFIC NAME: *Panthera tigris altaica*.
IDENTIFYING TRAITS: The Amur is the world's largest cat: it can attain a length of up to 14' from nose to tail tip (though 9'-10' is more typical) and reach a weight of 1,000 lbs (though 600-700 lbs is more usual). Its huge body is, in part, an adaptation to cool climates, for large bodies conserve heat better than small bodies.

Its (often white) "fur collar" is more pronounced than in other tiger subspecies, added protection against cold weather. Its heavy coat is lighter in color and more yellow-gold than other tiger subspecies (which tend to have darker orangish fur), its lower body and throat are generally more whitish than other subspecies, and its stripes are typically brownish rather than blackish. Males may grow a lion-like mane as they age.

DISTRIBUTION: Found mainly in the Russian Far East, which includes parts of Siberia, hence its alternate common name, the Siberian tiger. Its common name, the Amur tiger, comes from the fact that it also lives in Manchuria (a historical region of northeast China) along the Amur River, where temperature extremes range from far below zero in the winter to nearly 100 degrees Fahrenheit in the summer.

Some believe that small populations still exist in North Korea and central China. Currently there seems to be a thriving but small population of Amur tigers in Russia's Sikhote-Alin Mountains, where they live alongside other large carnivores like the leopard, the wolf, and the brown bear.

HABITAT: Grassland, forest, tundra, mountainous areas.

FEEDING AND HUNTING HABITS: A carnivorous, nocturnal, ambush-predator; will traverse hundreds of miles to locate game.

ENEMIES: Primarily humans.

POPULATION: There are probably less than 300 wild Amur tigers left. It is impossible to establish a precise number because, as with many other reclusive animals, the Amur often lives in remote areas that are inaccessible to researchers.

LIFE SPAN: About 10-15 years in the wild; as long as 20 years in captivity.

CONSERVATION STATUS: Endangered, with disease, poaching, logging, deforestation, habitat loss, and depletion of its prey base all pushing it toward extinction.

COMMENTS: The Amur's long thick coat permits it to endure temperatures as low as -49° Fahrenheit (that is, 49 degrees below zero), while additional fur on the paws helps insulate them from ice and snow. In general, this subspecies has a broader face and a huskier heavier body than most of its tiger cousins.

Efforts to maintain the purity of the Amur's bloodline are expedited through the Amur tiger SSP, or "Survival Species Plan," an accredited member-only program designed to oversee the population management, research, husbandry, and conservation of Amur tigers. Run by the Association of Zoos and Aquariums, this program forbids the breeding or registration of *hybrid* Amur tigers that could weaken the genes and lower the genetic diversity of this subspecies.

It is interesting to note that evolutionists have not been able to solve the mystery as to why the Amur tiger never developed a white camouflaging winter coat like other northern mammals—such as the snowshoe hare and the short-tailed weasel.

One school of thought holds that the Amur tiger is closely related to the extinct Caspian tiger, while another maintains that they are one and the same subspecies. If the latter turns out to be correct, the Amur tiger may eventually be used to repopulate the original territory of the Caspian tiger: Central and Western Asia.

BALI TIGER

2. COMMON NAME: **BALI TIGER**
ALTERNATE COMMON NAME: Balinese tiger.
SCIENTIFIC NAME: *Panthera tigris balica*.
IDENTIFYING TRAITS: At the time it flourished the Bali was the smallest member of the tiger family, and was thus diminutive compared to other tiger subspecies. About the size of a leopard, it has been estimated that males weighed around 220 lbs and grew up to about 7.5'; females probably weighed as much as 175 lbs while reaching a length of around 6.5'.

The Bali's thick fur was shorter and a darker yellowish-orange than other tigers and its white fur was whiter. In addition, its body possessed fewer and smaller stripes, which often wended around irregular black spots scattered over the body.
DISTRIBUTION: As its name indicates, this tiger was found solely on the little Indonesian island of Bali.
HABITAT: Jungle, grassland, forest.
FEEDING AND HUNTING HABITS: A carnivorous, nocturnal, ambush-predator.

ENEMIES: Primarily humans.
POPULATION: Zero.
LIFE SPAN: Probably about 10-12 years in the wild.
CONSERVATION STATUS: Extinct, due mainly to habitat loss and overhunting.
COMMENTS: Bali being a tiny island, it was inevitable that its tigers would eventually come into conflict with its ever increasing human population. After losing their fight with us they went the way of the dinosaur sometime between the late 1930s and 1940s.

It seems that no video footage or photographs were ever taken of a living Bali tiger before its extinction, and there are no known captive Bali tigers. Indeed, it appears that this subspecies was never kept by any zoo, perhaps further confirming its purported demise. Despite this, alleged sightings of the big cat continued into the 1970s.

This tiger may have been doomed from the start: always a relatively rare subspecies, the island of Bali is probably too small to provide food and shelter for a sizeable healthy breeding population of large carnivores. Even if it had endured into the present, there is no longer enough forest on Bali for it to survive in the wild.

Some hypothesize that the Bali tiger and the Javan tiger may be closely related, and that they divided into two separate subspecies after the last Ice Age, swimming to different islands as sea levels rose.

Since the Bali is now gone, however, there is no complete body to examine in order to determine its precise relationship to the Javan, or, for that matter, any other tiger subspecies. Only a few Bali tiger skulls, hides, and bones remain, preserved in various museum collections around the world. Debate over the taxonomy of this subspecies will continue.

The photo accompanying this entry is of a dead Bali tiger that was shot sometime in the 1920s.

BENGAL TIGER

☐ 3. COMMON NAME: **BENGAL TIGER**
ALTERNATE COMMON NAMES: Royal Bengal tiger, Indian tiger.
SCIENTIFIC NAME: *Panthera tigris tigris*.
ALTERNATE SCIENTIFIC NAME: *Panthera tigris bengalensis*.
IDENTIFYING TRAITS: The Bengal is the second largest of the tiger subspecies. Only the Amur tiger is bigger. The two, however, share similar pelage coloration and markings, for like the Amur's, the Bengal's coat is golden-orangish overlaid with dark stripes.

Male Bengal tigers attain an average weight of around 500 lbs (though large individuals may reach 850 lbs), and grow to an average length of between 9' and 10' from nose tip to tail tip; female Bengals reach an average weight of about 300 lbs (though 400 lbs is not unheard of), and grow to an average length of around 8.5' from nose tip to tail tip.

The Bengal tiger is responsible for two additional coat colorations, both of them unique in the tiger world: the famous white tiger (a result of a natural genetic mutation) and the black tiger (a pseudo-melanistic form that results from inbreeding). The white tiger has a whitish base coat with light gray or brownish stripes; the black tiger has a charcoal grayish base coat with off white or light yellowish stripes. Neither type is considered a distinct subspecies; they are merely variant mutated forms of the Bengal tiger.

DISTRIBUTION: The Bengal tiger is found in India, a land that is generally hot and humid, and where the temperature rarely drops below freezing. It is due to its presence here that it has earned its alternate common name, the Indian tiger. The Bengal also flourishes in Bhutan, Bangladesh, China, and Nepal.

HABITAT: Wet jungle, grassland, scrubland, humid rainforest, lowlands, dry deciduous woodlands, and mangrove swamps. Like other tigers, it requires an abundance of dense hiding cover for resting, reproduction, and hunting.

FEEDING AND HUNTING HABITS: A carnivorous, nocturnal, ambush-predator. The Bengal is known to attack, kill, and consume other tigers. An ardent hydrophile and an adept swimmer, it will pursue prey that attempts to flee into ponds, rivers, and lakes.

ENEMIES: Primarily humans.

POPULATION: Most estimates put its current wild population at between 2,000 and 4,000 individuals; although some believe there may be as few as 1,300 or as many as 5,000.

Whatever the precise number, it seems likely that there are more wild Bengals than the population of all other tiger subspecies combined. If true, it is the most common type of tiger.

LIFE SPAN: Estimates range from 8 years to 15 years in the wild; it may live for as long as 20 years in captivity.

CONSERVATION STATUS: Endangered.

COMMENTS: The national animal of both India and Bangladesh, the Bengal tiger has the distinction of possessing the largest and longest canines of any of the tiger subspecies, reaching a length of up to 4".

While the Bengal tiger seems to have the highest wild population of any of the remaining tiger subspecies, it is one of the most in demand for tiger parts and continues to be poached indiscriminately.

CASPIAN TIGER

4. COMMON NAME: **CASPIAN TIGER**
ALTERNATE COMMON NAMES: Mazandaran tiger, Persian tiger, Turan tiger, Turanian tiger, Hyrcanian tiger, road leopard, traveling leopard.
SCIENTIFIC NAME: *Panthera tigris virgata*.
IDENTIFYING TRAITS: Somewhat similar to the Bengal tiger in coloration and overall appearance, the Caspian had a thick husky body covered in long, dense, yellowish-brownish base fur.

The male grew to nearly 10' in total length and could attain a weight of around 500 lbs; the female reached a maximum length of 8.5' and an upper weight of around 300 lbs. This subspecies had large cheek ruffs, huge front limbs, broad front paws, massive claws, and smallish rounded ears. Its stripes were somewhat thin and ranged in shading from tan to dark brown.

Like most mammals, its pelage was countershaded: dark fur on the upper body, lighter fur on the lower body. The Caspian had the typical facial markings of the *Panthera tigris* group: a mixture of white, yellow-gold, and black stripes, with white and black eye rings.

DISTRIBUTION: The Caspian tiger was found throughout western and central Asia. More specifically it inhabited parts of Russia, Uzbekistan, Tajikistan, Mongolia, northern Iraq, Afghanistan, Turkey, China, and the

Caspian region (Turkmenistan, south Azerbaijan, and northern Iran), hence its common name.
HABITAT: Damp sparse forests, riparian woodlands, marshes, grassland, lake areas, muddy river banks, river deltas, arid deserts, wetlands, steppe, mangroves, thicketland, shrubland, and tugai (wet forests associated with flooding).
FEEDING AND HUNTING HABITS: A carnivorous, nocturnal, ambush-predator.
ENEMIES: Primarily humans.
POPULATION: Zero.
LIFE SPAN: Probably about 12 years in the wild.
CONSERVATION STATUS: Extinct, due to Russian extermination programs, overhunting, poaching, land reclamation projects, natural catastrophes, deforestation, farming, habitat loss, disease, and a decreasing prey base.
COMMENTS: Believed to be closely related to the Amur tiger, it is not known with any certainty when the Caspian tiger went extinct in the wild. Most estimates range from the 1950s to the 1990s; though at least two reliable authorities assert that the last one was killed in 1947. While none of these claims have been scientifically substantiated (and probably never will be), occasional sightings of the cat, along with its alleged pugmarks (paw prints), continue into the present day.

Several mounted specimens and a few skulls, skins, drawings, and photographs in museum collections are all that is left of the Caspian tiger. There are no known captives in zoos or circuses, and there seem to be none under private ownership.

Since some scientists consider the Caspian tiger not just related to, but actually identical to the Amur tiger, plans are being put into place to possibly repopulate the Caspian's original tugai-rich homeland (central Asia) with the Amur. These would come from zoo-bred specimens, orphaned cubs, and wild adult Amur tigers. Unfortunately, though still somewhat populous, the endangered Amur tiger is also in decline.

INDOCHINESE TIGER

☐ 5. COMMON NAME: **INDOCHINESE TIGER**
ALTERNATE COMMON NAMES: Corbett's tiger, Malayan tiger (a result of its confusion with the real Malayan tiger, *Panthera tigris jacksoni*).
SCIENTIFIC NAME: *Panthera tigris corbetti* (named after British hunter, naturalist, and conservationist James Edward "Jim" Corbett).
IDENTIFYING TRAITS: As an inhabitant of tropical areas, this tiger is relatively small (an evolutionary adaptation to hot climates). It has a richer, darker golden-orange base coat, narrower and shorter stripes, and more stripes, than other tiger subspecies. While females may attain a nose-to-tail tip length of 8' and a weight of 250 lbs, males can reach 9' and a weight of up to 400 lbs.
DISTRIBUTION: The Indochinese tiger is a native of China, mainly from the southeastern part of the country. A wide roaming big cat with a broad range, it is also found in Vietnam, Thailand, Cambodia, Malaysia, Laos, and Myanmar.
HABITAT: Secluded mountains, hills, humid forests, remote wilderness

areas.

FEEDING AND HUNTING HABITS: A carnivorous, nocturnal, ambush-predator.

ENEMIES: Primarily humans.

POPULATION: Due to the inaccessible habitat it prefers, its wild population is unknown. There may be as few as 300 or as many as 1,500 individuals left in various wilderness areas.

LIFE SPAN: Probably up to 15 years or so in the wild; as long as 25 years in captivity.

CONSERVATION STATUS: Endangered.

COMMENTS: This tiger was once thought to be identical with what is now called the Malayan tiger (*Panthera tigris jacksoni*). Recent genetic analysis has determined that they are two separate but closely related subspecies. Several authorities, however, continue to maintain that they are a single subspecies, *Panthera tigris corbetti*. Taxonomic discussions have yet to lead to scientific consensus.

A well-known man-eater, the Indochinese tiger has developed the self-destructive habit of haunting villages and attacking people. Presently there are several dozen individuals living in zoos, parks, and reserves that could be used for breeding. However, due to uncertainties concerning the genetic backgrounds of these captive specimens, this subspecies may be beyond rehabilitation. At the current reduction rate, some predict that the Indochinese tiger will soon be extinct in the wild.

JAVAN TIGER

6. COMMON NAME: **JAVAN TIGER**
SCIENTIFIC NAME: *Panthera tigris sondaica.*
IDENTIFYING TRAITS: A smallish tiger that is sometimes compared in overall appearance to the Sumatran tiger, the male Javan grew to a length of around 8' and weighed an average of 250 lbs. Its snout was narrower and its coat was a deeper orange with thinner stripes than most other subspecies.

Despite its moderate size, it had the longest facial vibrissae of any tiger. Its stripes could be quite irregular, pale, and even absent on some parts of the body, with spots replacing them altogether in some areas.
DISTRIBUTION: As its name indicates, this big cat was once found on the Indonesian island of Java. At some time in the distant past it probably reached Java by swimming, perhaps via Malaysia and Sumatra. Fossil evidence indicates that the Javan tiger may have also once inhabited the island of Borneo as well as the island of Palawan in the Philippines.
HABITAT: Mainly forests.
FEEDING AND HUNTING HABITS: A carnivorous, nocturnal, ambush-predator.
ENEMIES: Primarily humans.
POPULATION: Zero.

LIFE SPAN: Probably 10-14 years or so in the wild.

CONSERVATION STATUS: Extinct, due to human population growth, overhunting, deforestation, plantation farming, habitat loss, and a decreasing prey base.

COMMENTS: The Javan tiger was once highly revered by Hindus and played an integral role in their religion. Early folklore held that the body of the Javan tiger was the abode of reincarnated ancestors, giving it a heightened sacrality among believers.

This subspecies was very common into the early 20th Century, so much so that it was considered a pest. It began fading from the scene in the 1940s as we gradually supplanted it across its range. Deforestation and conflicts with farmers and ranchers pushed it further toward oblivion.

Java's largest animal sanctuary, Ujong Kulon, possessed several individuals into the 1960s, while the last known wild Javan tiger was spotted in 1976. However, pugmarks were seen in 1979, after which the animal seems to have vanished. This indicates that it probably finally went extinct sometime in the 1980s. There are no known captive Javan tigers, though several photographs, skins, skulls, and skeletons are held in museum collections.

Despite the assumptions of mainstream science, alleged sightings (mainly by native peoples) are still being reported, thus it is possible that a few members of this subspecies continue to thrive in the remote wilderness. And indeed, some conservation groups believe it may still be alive. Unfortunately, though hopeful minds remain open, there is currently no hard evidence to sustain this view.

MALAYAN TIGER

☐ 7. COMMON NAME: **MALAYAN TIGER**
ALTERNATE COMMON NAME: Harimau Belang (a moniker used by Malaysian peoples).
SCIENTIFIC NAME: *Panthera tigris jacksoni* (named after British author and tiger conservationist, Peter Jackson).
ALTERNATE SCIENTIFIC NAME: *Panthera tigris malayensis* (known by this trinomial within Malay).
IDENTIFYING TRAITS: The Malayan tiger is one of the smaller of the tiger subspecies, with the male growing from 6.5' to 9' in length (though 7.5' is average) and attaining an upper weight of 300 lbs. The female reaches a length of between 5.5' and 7.5', and weighs from 175 lbs to 225 lbs.

This smallish big cat sports typical tiger traits and coloration: a powerful lithe body; a yellowish-orange base coat with thinnish, well defined, black vertical stripes; white patches over the face, head, chest, lower body, and inner legs; small rounded erect ears; long stiff vibrissae; and large round yellow-green eyes.
DISTRIBUTION: A strictly Malaysian cat, it is found only on the Malay Peninsula and in southern Thailand.
HABITAT: Dense tropical and subtropical forests. A riparian mammal, it is an excellent swimmer that prefers living near streams and rivers.
FEEDING AND HUNTING HABITS: A carnivorous, nocturnal,

ambush-predator. It can attain a speed of up to 40 mph for short distances. As the accompanying color photo shows, this tiger enjoys playing in water and will swim merely for the joy of it; but it will also enter water in pursuit of prey.

ENEMIES: Primarily humans.

POPULATION: There are probably only between 250 and 300 individual wild Malayan tigers left, making it one of the world's rarest cats.

LIFE SPAN: From 15-20 years in the wild, though some individuals may live as long as 25 years.

CONSERVATION STATUS: Critically endangered.

COMMENTS: At one time it was believed that the Malayan tiger (*Panthera tigris jacksoni*) and the Indochinese tiger (*Panthera tigris corbetti*) were the same subspecies. Though closely related, recent genetic studies reveal that they are different and therefore two distinct subspecies. Despite this, some authorities continue to maintain that—since they lack clear morphological differences—the two are identical, placing both under the name *Panthera tigris corbetti*. Still others accept provisional status for *Panthera tigris jacksoni*. Taxonomic review and debate is ongoing.

The national animal of Malaysia (whose feline image appears on its coat of arms), like all other wild tigers the Malayan is precipitously dying out due to a myriad of pressures, from logging, farming, overhunting, and poaching (for body parts), to habitat loss, conflicts with local communities, deforestation, and an ever decreasing prey base. Breeding programs in Malaysia are having some success in producing healthy cubs, but the danger of extinction remains very real.

The national animal of Malaysia, *Panthera tigris jacksoni* is featured prominently on the country's coat of arms.

NGANDONG TIGER

8. COMMON NAME: **NGANDONG TIGER**
SCIENTIFIC NAME: *Panthera tigris soloensis*.
IDENTIFYING TRAITS: The Ngandong tiger is only known from a handful of fossils, which reveal that it was one of the largest felines to have ever lived, with a probable resemblance to the modern Bengal tiger in size and appearance. It pelage coloration and markings are, of course, unknown, but it may have looked something like the *Panthera tigris* pictured above.

Males probably grew to about 7' (from head to rump), with a total length of 12' (from nose tip to tail tip), and an estimated weight of between 800 lbs and 1,000 lbs. Females were likely half the size of the males in weight and other dimensions.
DISTRIBUTION: Predominately the Sundaland region of Indonesia.
HABITAT: Tropical rain forests.
FEEDING AND HUNTING HABITS: Probably the same as modern living tigers; that is, it was most likely a carnivorous, nocturnal, ambush-predator.
ENEMIES: Other carnivores, and possibly prehistoric humans, such as *Homo erectus*.
POPULATION: Zero.

LIFE SPAN: Unknown, but probably similar to living tigers: around 10-15 years in the wild.

CONSERVATION STATUS: Extinct.

COMMENTS: Remains of the Ngandong tiger were unearthed around Ngandong, China, which gives it its common name. This gigantic felid flourished during the Pleistocene Epoch, some 100,000 years ago. We do not know why this subspecies became extinct, but possibilities range from a great natural disaster to overhunting by early humans.

SOUTH CHINA TIGER

☐ 9. COMMON NAME: **SOUTH CHINA TIGER**
ALTERNATE COMMON NAMES: Amoy tiger, Xiamen tiger.
SCIENTIFIC NAME: *Panthera tigris amoyensis*.
IDENTIFYING TRAITS: The South China tiger is the smallest of the living tiger subspecies: the male may grow from 7.5' to 8.5' in length and attain a weight of about 375 lbs; the female can grow up to 7.5' in length and reach a weight of about 230 lbs or so. This tiger's base coat is darker (a rich chocolate-orange) and its stripes are set more closely together than in other subspecies. Notably, its tail ends in a stub rather than a gradually narrowing point.
DISTRIBUTION: As its name suggests, the South China tiger is found in southern China, and more specifically in the central provinces of Jiangxi, Hunan, Fujian, and Guangdong.
HABITAT: Forests, caves, rocky situations, and mountainous regions.
FEEDING AND HUNTING HABITS: A carnivorous, nocturnal, ambush-predator.
ENEMIES: Primarily humans.
POPULATION: There are probably only 50-100 wild South China tigers left; some maintain that there are even less, perhaps as few as 20. Still

others believe that it is already extinct in the wild, for none have been observed in their native habitat for over 20 years. Fortunately for this small-bodied subspecies, between 50 and 100 individuals are being kept in Chinese zoos.

LIFE SPAN: Probably as long as 15 years in the wild; perhaps 20 years in captivity.

CONSERVATION STATUS: Critically endangered; the last known wild specimen was killed in 1994, making it functionally extinct in the wild.

COMMENTS: Based on several of its primitive-like features (such as its short thin skull and reduced pupillary distance), it is possible that the South China tiger is the predecessor of all other tiger subspecies. If correct, then the other living and recently extinct tigers evolved from the South China tiger. These forms then dispersed out across Asia and, in response to the demands of their particular habitats, developed into the other eleven currently recognized subspecies.

Though some believe there still may be a few individual South China tigers remaining in the wild, this seems doubtful due to lack of sightings. Even if there were, by now the wild population would be too small to sustain itself, and viruses and natural disasters would pose a constant threat to the entire group.

As noted, a number of individuals remain in zoos. However, many of these are not breedable (for various reasons, such as inferior genetics and old age). This means that currently, of all the tiger subspecies, the South China is the most liable to soon become fully extinct; that is, both in the wild and in captivity.

SUMATRAN TIGER

☐ 10. COMMON NAME: **SUMATRAN TIGER**
SCIENTIFIC NAME: *Panthera tigris sumatrae*.
IDENTIFYING TRAITS: Though it is the smallest of the living tiger subspecies, it is the largest carnivore on its island home. Its rich yellowish-orange coat is overlaid with double black stripes set close together. It has striped front limbs and possesses one of the darkest base coats of any tiger. Its prominent mane, cheek ruffs, eye rings, and beard are particularly diagnostic. Males can reach a total length of 8' and a weight of about 250 lbs, while females can attain a total length of 7.5' and a weight of around 200 lbs.
DISTRIBUTION: The Sumatran tiger is found only in Indonesia, and more specifically on the island of Sumatra, hence both its common and

scientific names. It may have reached Sumatra by swimming there sometime in the distant past.

HABITAT: Pine forests, volcanic mountainous terrain, peat swamps, woodlands, and tropical rainforests.

FEEDING AND HUNTING HABITS: This carnivore is a nocturnal ambush-predator. It is fond of water and will swim after prey using its powerful broad paws and webbed toes. It can reach a speed of 40 mph for a few seconds, allowing it to catch all manner of prey.

ENEMIES: Primarily humans.

POPULATION: Estimates range from 400 to 600 individuals in the wild.

LIFE SPAN: May live as long as 15 years or more in the wild; as long as 25 years in captivity.

CONSERVATION STATUS: Critically endangered, due mainly to overhunting, poaching, forest fires, lack of conservation awareness in villages, and habitat loss from logging and human population growth.

COMMENTS: Since it has been isolated on its island home for at least 10,000 years, the Sumatran tiger may be evolving into a separate and distinct species. If true, conservation efforts should be doubly increased to save this unique big cat from extinction. There are a number of captive specimens in zoos around the world that may eventually assist in that very effort.

TRINIL TIGER

11. COMMON NAME: **TRINIL TIGER**
ALTERNATE COMMON NAME: Pleistocene tiger.
SCIENTIFIC NAME: *Panthera tigris trinilensis*.
IDENTIFYING TRAITS: A smallish tiger, the male grew to 6.25' in length, stood a little over 3' in height, and weighed approximately 330 lbs. Its pelage coloration and markings are unknown, but may have been similar to today's tigers, as the inventive illustration above infers.
DISTRIBUTION: Fossil evidence of the Trinil tiger was found at the Trinil paleontology site on the island of Java, Indonesia, hence its common and scientific names.
HABITAT: Probably typical tiger land: forests, mangrove swamps, grassland, and river banks.
FEEDING AND HUNTING HABITS: Likely the same as modern living tigers: a carnivorous, nocturnal, ambush-predator.
ENEMIES: Other carnivores, and possibly prehistoric hominids.
POPULATION: Zero.
LIFE SPAN: Unknown, but assumed to be similar to living tigers; that is, it may have had a life expectancy of around 10-15 years in the wild.
CONSERVATION STATUS: Extinct (for reasons unknown, possibly major climatological and geological events).

COMMENTS: The Trinil tiger lived during the Pleistocene Epoch, sometime between 1.3 million years and 700,000 years ago. The precise time of its extinction remains a mystery, but it seems likely to have occurred sometime during the Middle Paleolithic Period (between 150,000 and 40,000 years ago). Owing to a paucity of fossil evidence, we have very little definitive knowledge about this extinct prehistoric big cat.

WANHSIEN TIGER

12. COMMON NAME: **WANHSIEN TIGER**
ALTERNATE COMMON NAME: Antique tiger.
SCIENTIFIC NAME: *Panthera tigris acutidens*.
IDENTIFYING TRAITS: A large subspecies, its body grew to about 7' in length; from nose tip to tail tip it may have reached 10'. It stood about 3.5' at the shoulder and may have weighed approximately 750 lbs.
DISTRIBUTION: Predominately eastern China, but also much of Asia.
HABITAT: Pine forests and subtropical forests.
FEEDING AND HUNTING HABITS: Likely the same as modern living tigers: a carnivorous, nocturnal, ambush-predator.
ENEMIES: Other carnivores.
POPULATION: Zero.
LIFE SPAN: Unknown, but probably similar to living tigers; that is, it may have had a life expectancy of around 10-12 years in the wild.
CONSERVATION STATUS: Extinct.
COMMENTS: Remains of the Wanhsien tiger were discovered in the Wanhsien District of China, hence its common name. It flourished during the Pleistocene Epoch, about 75,000 years ago. It is the earliest known extinct tiger subspecies. Not enough of its remains have been found to develop a complete image of its appearance, but it may have looked something like the South China tiger (pictured above)—which some believe may be its descendant.

The End

A mother tiger protecting her sleeping cubs from a snake.

BIBLIOGRAPHY

And Suggested Reading

Academic American Encyclopedia. New York: Grolier Academic Reference, 1998.

Adams, William Henry Davenport. *Animal Life Throughout the Globe: An Illustrated Book of Natural History*. London, UK: T. Nelson and Sons, 1876.

Agassiz, Louis. *An Introduction to the Study of Natural History*. New York: Greeley and McElrath, 1847.

——. *Contributions to the Natural History of the United States of America*. Boston, MA: Little, Brown and Co., 1860.

Agustí, Jordi, and Mauricio Antón. *Mammoths, Sabertooths, and Hominids: 65 Million Years of Evolution in Europe*. New York: Columbia University Press, 2002.

Ahern, Albert M. *Fur Facts: A Book of Knowledge*. St. Louis, MO: C. P. Curran, 1922.

Allen, Glover M. *The Mammals of China and Mongolia: Natural History of Central Asia*. New York: American Museum of Natural History, 1938-1940.

——. *Extinct and Vanishing Mammals of the Western Hemisphere*. American Committee for International Wildlife Protection, 1942.

Anderson, Sydney, and J. Knox Jones. *Orders and Families of Recent Mammals of the World*. New York: John Wiley and Sons, 1984.

Andrews, Roy Chapman, and Yvette Borup Andrews. *Camps and Trails in China: Narrative of Exploration, Adventure, and Sport in Little-known China*. New York: D. Appleton and Co., 1919.

Applewhite, Ashton, William R. Evans III, and Andrew Frothingham (eds.). *And I Quote: The Definitive Collection of Quotes, Sayings, and Jokes for the Contemporary Speechmaker*. New York: St. Martin's Press, 2003.

Asdell, Sydney Arthur. *Patterns of Mammalian Reproduction*. Ithaca, NY: Comstock Publishing, 1964.

Attenborough, David. *Life on Earth: A Natural History*. New York: Little, Brown and Co., 1981.

———. *The Living Planet: A Portrait of the Earth.* New York: Little, Brown and Co., 1984.

———. *The Life of Mammals.* London, UK: BBC Books, 2002.

Balfour, Edward. *The Cyclopedia of India and of Eastern and Southern Asia: Commercial, Industrial, and Scientific.* 3 vols. London, UK: Bernard Quaritch, 1885.

Ball, Valentine. *Jungle Life in India; or the Journeys and Journals of an Indian Geologist.* London, UK: Thomas de la Rue and Co., 1880.

Barras, Julius. *India and Tiger-Hunting.* 2 vols. London, UK: Swan Sonnenschein and Co., 1885.

Becker, John E. *Wild Cats: Past and Present.* Minneapolis, MN: Millbrook Press, 2008.

Beeton, Samuel Orchart. *Beeton's Dictionary of Natural History: A Compendious Encyclopedia of the Animal Kingdom.* London, UK: Ward, Lock, and Tyler, 1871.

Beilby, Ralph. *A General History of Quadrupeds.* New Castle Upon Tyne, UK: T. Bewick and S. Hodgson, 1807.

Bellani, Giovanni Giuseppe. *Felines of the World: Discovery in Taxonomic Classification and History.* London, UK: Academic Press, 2020.

Beolens, Bo, Michael Watkins, and Michael Grayson. *The Eponym Dictionary of Mammals.* Baltimore, MD: Johns Hopkins University Press, 2009.

Berger, P. *William Blake: Poet and Mystic.* New York: E. P. Dutton and Co., 1915.

Biedermann, Hans. *Dictionary of Symbolism: Cultural Icons and the Meanings Behind Them.* 1989. New York: Facts on File, 1992 ed.

Boreman, Thomas. *A Description of 300 Animals.* London, UK: self-published, 1769.

Borror, Donald Joyce. *Dictionary of Word Roots and Combining Forms: Compiled from the Greek, Latin, and other Languages, With Special Reference to Biological Terms and Scientific Names.* 2 vols. Mountain View, CA: Mayfield, 1960.

Bortolotti, Dan. *Tiger Rescue: Changing the Future for Endangered Wildlife.* Buffalo, NY: Firefly Books 2003.

Bourbeau, André-François. *Wilderness Secrets Revealed: Adventures of a Survivor.* Toronto, CAN: Dundurn, 2013.

Box, Hilary O., and Kathleen R. Gibson (eds.). *Mammalian Social Learning: Comparative and Ecological Perspectives.* Cambridge, UK: Cambridge

University Press, 1999.
Boyce, Mark S. (ed.). *Evolution of Life Histories of Mammals: Theory and Pattern*. New Haven, CT: Yale University Press, 1988.
Bourlière, François. *The Natural History of Mammals*. New York: Alfred A. Knopf, 1954.
Brakefield, Tom. *Big Cats: Kingdom of Might*. St. Paul, MN: Voyageur Press, 1993.
Brehm, Alfred Edmund. *Brehm's Life of Animals: A Complete Natural History for Popular Home Instruction and for the Use of Schools*. Chicago, IL: A. N. Marquis and Co., 1896.
Buchanan, Daniel Crump. *Japanese Proverbs and Sayings*. Norman, OK: University of Oklahoma Press, 1965.
Burton, Maurice, and Robert Burton. *International Wildlife Encyclopedia*. Tarrytown, NY: Marshall Cavendish, 2002.
Caldwell, Harry R. *Blue Tiger*. New York: Abingdon Press, 1924.
Carrington, Richard. *The Mammals*. New York: Time-Life, 1965.
Carroll, Robert L. *Vertebrate Paleontology and Evolution*. New York: W. H. Freeman, 1990.
Carson, Rachel. *Silent Spring*. Boston, MA: Houghton Mifflin Co., 1962.
——. *The Sense of Wonder*. New York: Harper and Row, 1965.
Chamberlain, Jacob. *In the Tiger Jungle*. New York: Fleming H. Revell Co., 1896.
Chambers's Encyclopaedia: A Dictionary of Universal Knowledge for the People. Edinburgh, Scotland: Chambers Publishing Company, 1860.
Chambers, Robert. *Vestiges of the Natural History of Creation*. New York: William H. Colyer, 1846.
Chinsamy-Turan, Anusuya. *Forerunner of Mammals: Radiation, Histology, Biology*. Bloomington, IN: Indiana University Press, 2012.
Christiansen, Per. *The Encyclopedia of Animals*. London, UK: Amber Books, 2006.
Chung, Vinh (with Tim Downs). *Where the Wind Leads: A Refugee Family's Miraculous Story of Loss, Rescue, and Redemption*. Nashville, TN: Thomas Nelson, 2014.
Cirlot, J. E. *A Dictionary of Symbols*. New York: Philosophical Library, 1962.
Compton's Pictured Encyclopedia. Chicago, IL: F. E. Compton and Co., 1929.
Cooper, Courtney Ryley. *Lions 'N' Tigers 'N' Everything*. Boston, MA: Little, Brown, and Co., 1924.

Cope, Edward D. "On the Extinct Cats of America." *The American Naturalist*, Vol. 14, No. 12, December 1880.

Corbet, Gordon Barclay. *A World List of Mammalian Species*. London, UK: Natural History Museum Publications, 1991.

Corbett, Jim. *Man-Eaters of Kumaon*. Oxford, UK: Oxford University Press, 1944.

Cornish, Charles J. *Wild Animals in Captivity*. New York: Macmillian and Co., 1894.

——. (ed.). *Mammals of Other Lands*. New York: The University Society, 1917.

Covarrubias, Miguel. *Island of Bali*. New York: Alfred A. Knopf, 1937.

Crooke, William. *An Introduction to the Popular Religion and Folklore of Northern India*. Allahabad, India: Government Press, 1894.

——. *Popular Religion and Folklore in Northern India*. 2 vols. 1896. Reprint, Delhi, India: Munshiram Manoharlal, 1968.

Cuvier, Georges. *The Animal Kingdom: Arranged in Conformity to its Organization*. New York: G. and C. and H. Carvill, 1831.

Cuvier, Georges, et al. *A System of Natural History; Containing Scientific and Popular Descriptions of Various Animals*. Brattleboro, VT: Peck and Wood, 1834.

Daithota, Shilpa Bhat. *Indians in Victorian Children's Narratives: Animalizing the Native, 1830-1930*. Lanham, MD: Lexington Books, 2017.

Daniels, Cora Linn, and C. M. Stevans (eds.). *Encyclopedia of Superstitions, Folklore, and the Occult Sciences of the World*. 3 vols. Milwaukee, WI: J. H. Yewdale and Sons Co., 1903.

Davis, Buddy, and Kay Davis. *Magnificent Mammals: Marvels of Creation*. Green Forest, AR: Master Books, 2006.

Davis, David Edward, and Frank B. Golley. *Principles in Mammalogy*. New York: Reinhold, 1963.

Davis, J. R. Ainsworth. *The Natural History of Animals: The Animal Life of the World in its Various Aspects and Relations*. London, UK: Gresham Publishing Co., 1905.

DeBlase, Anthony F. *A Manual of Mammalogy: With Keys to Families of the World*. Dubuque, IA: William C. Brown, 1980.

Denis, Armand. *Cats of the World*. New York: Houghton Mifflin, 1964.

Die Gartenlaube: Illustrirtes Familienblatt (The Garden Arbor: Illustrated Family Journal). Leipzig, Germany: Ernst Keil, 1880.

Douglas, Carstairs. *Chinese-English Dictionary of the Vernacular or Spoken*

Language of Amoy. London, UK: Presbyterian Church of England, 1899.

Drew, Liam. *I Mammal: The Story of What Makes Us Mammals*. London, UK: Bloomsbury, 2017.

Duff, Andrew, and Ann Lawson. *Mammals of the World: A Checklist*. New Haven, CT: Yale University Press, 2004.

Duffy, Andrea E. *Nomad's Land: Pastoralism and French Environmental Policy in the Nineteenth-century Mediterranean World*. Lincoln, NE: University of Nebraska Press, 2019.

Durrell, Lee. *State of the Ark*. London, UK: Gaia Books, 1986.

Eisenberg, John F. *The Mammalian Radiations: An Analysis of Trends in Evolution, Adaptation, and Behavior*. Chicago, IL: University of Chicago Press, 1983.

Encyclopedia Britannica, or, a Dictionary of Arts and Sciences, Compiled Upon a New Plan. 1768. London, UK: Encyclopedia Britannica, Inc., 1771 ed.

English Cyclopaedia. London, UK: Bradbury, Evans, and Co., 1867.

English, Douglas. *Beasties Courageous: Studies of Animal Life and Character*. London, UK: S. H. Bousfield and Co., 1905.

Entwistle, Abigail, and Nigel Dunstone (eds.). *Priorities for the Conservation of Mammalian Diversity: Has the Panda Had its Day?* Cambridge, UK: Cambridge University Press, 2000.

Everyman's Encyclopaedia. London, UK: J. M. Dent and Sons, 1958.

Ewer, R. F. *The Carnivores*. Ithaca, NY: Comstock Publishing, 1986.

Fayrer, Joseph. *The Royal Tiger of Bengal: His Life and Death*. London, UK: J. and A. Churchill, 1875.

Feldhammer, George A., Lee C. Drickamer, Stephen H. Vessey, Joseph F. Merritt, and Carey Krajewski. *Mammalogy: Adaptation, Diversity, Ecology*. Baltimore, MD: Johns Hopkins University Press, 2015.

Firouz, E. *A Guide to the Fauna of Iran*. Tehran, Iran: Iran University Press, 2000.

Fitzinger, Leopold Joseph. *Bilder-Atlas zur Wissenschaftlich-Populären Naturgeschichte der Säugethiere in Ihren Sämmtlichen Hauptformen*. Vienna, Austria: n.p., 1860.

Forrest, George. *Cities of India*. Westminster, UK: Archibald Constable and Co., 1903.

Fowler, Charles W., and Tim D. Smith (eds.). *Dynamics of Large Mammal Populations*. Caldwell, NJ: Blackburn Press, 2004.

Fryxell, John M., Anthony R. E. Sinclair, and Graeme Caughley. *Wildlife*

Ecology, Conservation, and Management. 1994. Oxford, UK: John Wiley and Sons, 2014 ed.

Funk and Wagnalls New Standard Encyclopedia of Universal Knowledge. New York: Funk and Wagnalls, 1931.

Gauvin, Marshall J. *The Illustrated Story of Evolution.* New York: Peter Eckler, 1921.

Gibson, David M., and Robert A. Harris. *Metabolic Regulation in Mammals.* London, UK: Taylor and Francis, 2002.

Goldsmith, Oliver. *Goldsmith's History of Man and Quadrupeds.* 2 vols. London, UK: Smith, Elder and Co., 1838.

Goodrich, Samuel G. *Parley's Book of Quadrupeds: for Youth.* New York: R. T. Young, 1853.

———. *A Pictorial Natural History: Embracing a View of the Mineral, Vegetable, and Animal Kingdoms.* Boston, MA: James Munroe and Co., 1854.

———. *Illustrated Natural History of the Animal Kingdom.* New York: Derby and Jackson, 1859.

Gössling, Stefan, and C. Michael Hall (eds.). *Tourism and Global Environmental Change: Ecological, Social, Economic, and Political Interrelationships.* Oxford, UK: Routledge, 2006.

Goswami, Anjali, and Anthony Friscia (eds.). *Carnivoran Evolution: New Views on Phylogeny, Form and Function.* Cambridge, UK: Cambridge University Press, 2010.

Gotch, Arthur Frederick. *Mammals - Their Latin Names Explained: A Guide to Animal Classification.* London, UK: Blandford Press, 1979.

Gouldsbury, C. E. *Tiger Slayer by Order.* New York: E. P. Dutton and Co., 1915.

Grant, James. *Cassell's Illustrated History of India.* London, UK: Cassell, Petter, and Galpin, 1876.

Gray, Henry. *Anatomy of the Human Body.* Philadelphia, PA: Lea and Febiger, 1918.

Grzimek, Bernhard. *Grzimek's Encyclopedia: Mammals.* New York: McGraw-Hill, 1989.

Gubernick, David J., and Peter H. Klopfer. *Parental Care in Mammals.* New York: Plenum Press, 1981.

Guggisberg, Charles Albert Walter. *Wild Cats of the World.* New York: Taplinger Publishing, 1975.

Gunderson, Harvey L. *Mammalogy.* New York: McGraw-Hill, 1976.

Gupta, Om. *Encyclopedia of India, Pakistan and Bangladesh.* (9 vols.) Delhi,

India: Isha Books, 2006.
Gutteridge, Anne C. *Barnes and Noble Thesaurus of Biology: The Principles of Biology Explained and Illustrated.* New York: Barnes and Noble Books, 1983.
Hagenbeck, Carl. *Beasts and Men: Being Carl Hagenbeck's Experiences for a Half Century Among Wild Animals.* London, UK: Longmans, Green, and Co., 1910.
Hallett, Mark, and John M. Harris. *On the Prowl: In Search of Big Cat Origins.* New York: Columbia University Press, 2020.
Hanel, Rachael. *Tigers.* Mankato, MN: Creative Education, 2009.
Harmsworth's Universal Encyclopedia. London, UK: The Educational Book Co., 1921.
Harris, Larry D. *The Fragmented Forest: Island Biogeography and the Preservation of Biotic Diversity.* Chicago, IL: University of Chicago Press, 1984.
Hawksworth, David L., and Alan t. Bull (eds.). *Vertebrate Conservation and Biodiversity.* Dordrecht, The Netherlands: Springer, 2007.
Hayssen, Virginia, and Teri Orr. *Reproduction in Mammals: The Female Perspective.* Baltimore, MD: Johns Hopkins University Press, 2017.
Hayssen, Virginia, Ari Van Tienhoven, and Ans van Tienhoven. *Asdell's Patterns of Mammalian Reproduction: A Compendium of Species-Specific Data.* 1946. Ithaca, NY: Comstock, 1993 ed.
Heiberg, Arthur B. (ed.). *Illustrated World* (Vol. 38, No. 1). Chicago, IL: R. T. Miller Jr., 1922.
Hill, Richard W., Gordon A. Wyse, and Margaret Anderson. *Animal Physiology.* Oxford, UK: Oxford University Press, 2016 ed.
Hillard, Darla. *Vanishing Tracks.* New York: William Morrow, 1989.
Hillier, S. H., D. W. H. Walton, and D. A. Wells (eds.). *Calcareous Grasslands: Ecology and Management.* Bluntisham, UK: Bluntisham Books, 1990.
Holder, Charles Frederick. *Half Hours With the Mammals.* New York: American Book Co., 1907.
Holmes, Martha, and Michael Gunton. *Life: Extraordinary Animals, Extreme Behaviour.* London, UK: BBC Books, 2009.
Honacki, James H., Kenneth E. Kinman, and James W. Koeppl (eds.). *Mammal Species of the World: A Taxonomic and Geographic Reference.* Washington, D.C.: Natural Science Collections Alliance, 1982.
Hoogerwerf, Andries. *Udjung Kulon: The Land of the Last Javan Rhinoceros.*

Leiden, Germany: Brill, 1970.

Horne, Thomas Hartwell. *Landscape Illustrations of the Bible*. 2 vols. London, UK: John Murray, 1836.

Hornocker, Maurice, and Sharon Negri (eds.) *Cougar: Conservation and Ecology*. Chicago, IL: University of Chicago Press, 2009.

Howell, F. Clark. *Early Man*. New York: Time-Life Books, 1965.

Howitt, Mary. *Sketches of Natural History*. London, UK: Effingham Wilson, 1834.

———. *Mary Howitt's Illustrated Library for the Young*. London, UK: W. Kent and Co., 1855.

Hulme, F. Edward. *Natural History: Lore and Legend*. London, UK: Bernard Quaritch, 1895.

Hunter, John. *Essays and Observations on Natural History, Anatomy, Physiology, Psychology, and Geology*. London, UK: John Van Voorst, 1861.

Hunter, Luke. *Carnivores of the World*. 2011. Princeton, NJ: Princeton University Press, 2018 ed.

———. *Wild Cats of the World*. London, UK: Bloomsbury, 2015.

Hutyra, Franz, and Josef Marek. *Special Pathology and Therapeutics of the Diseases of Domestic Animals*. Chicago, IL: Alexander Eger, 1912.

Huxley, Thomas Henry. *Man's Place in Nature, and Other Anthropological Essays*. London, UK: Macmillan and Co., 1894.

Jackson, A. V. Williams (ed.). *History of India*. London, UK: Grolier Society, 1907.

Jackson, Peter. *Endangered Species: Tigers*. New York: Chartwell Books, 1990.

Jaeger, Edmund C. *A Source-Book of Biological Names and Terms*. Springfield, IL: Charles C. Thomas, 1950.

Jaeger, Ellsworth. *Tracks and Trailcraft*. New York: Macmillan, 1948.

Jalais, Annu. *Forest of Tigers: People, Politics and Environment in the Sundarbans*. London, UK: Routledge, 2010.

Jardine, William (ed.). *The Natural History of the Felinae*. Edinburgh, Scotland: W. H. Lizars, 1834.

———. *The Naturalist's Library*. London, UK: Henry G. Bohn, 1866.

Jenyns, Leonard. *Observations in Natural History*. London, UK: John Van Voorst, 1846.

Jerdon, Thomas C. *The Mammals of India: A Natural History of all the Animals Known to Inhabit Continental India*. London, UK: John Wheldon, 1874.

Johnston, Alexander Keith. *School Atlas of Physical Geography*. Edinburgh,

Scotland: William Blackwood and Sons, 1873.
Jones, Stephen. *The Natural History of Beasts, Compiled From the Best Authorities*. London, UK: self-published, 1793.
Jo, Yeong-Seok, John T. Baccus, and John L. Koprowski. *Mammals of Korea*. Incheon, Korea: National Institute of Biological Resources, 2018.
Journal of Mammalogy. Baltimore, MD: American Society of Mammalogists, 1919-present.
Karanth, K. Ullas. *The Way of the Tiger: Natural History and Conservation of the Endangered Big Cat*. Stillwater, MN: Voyageur Press, 2001.
———. (ed.) *Tiger Tales: Tracking the Big Cat Across Asia*. New Delhi, India: Penguin Books, 2006.
Kemp, T. S. *The Origin and Evolution of Mammals*. Oxford, UK: Oxford University Press, 2005.
Kerr, Robert. *A General History and Collection of Voyages and Travels*. Edinburgh, Scotland: selfpublished, 1815.
Kielan-Jaworowska, Zofia, Richard L. Cifelli, and Zhe-Xi Luo. *Mammals From the Age of Dinosaurs: Origins, Evolution, and Structure*. New York: Columbia University Press, 2004.
Kingsley, John Sterling. *The Riverside Natural History: Vol. 5, Mammals*. London, UK: Kegan Paul, Trench and Co., 1888.
Kipling, Rudyard. *The Jungle Book*. New York: The Century Co., 1920.
Kitchener, Andrew. *The Natural History of the Wild Cats*. Ithaca, NY: Comstock Publishing, 1991.
Knight, Charles. *Natural History*. London, UK: Bradbury, Evans, and Co., 1866.
Knox, Thomas W. *Overland Through Asia: Pictures of Siberian, Chinese, and Tartar Life*. Hartford, CT: American Publishing Co., 1870.
Lanz, Tobias J. *The Life and Fate of the Indian Tiger*. Santa Barbara, CA: ABC-Clio, 2009.
Lawlor, Timothy E. *Handbook to the Orders and Families of Living Mammals*. Eureka, CA: Mad River Press, 1979.
Lawrence, W. *Lectures on Physiology, Zoology, and the Natural History of Man*. London, UK: selfpublished, 1819.
Le Clerc, George Louis. *A Natural History, General and Particular; Containing the History and Theory of the Earth, a General History of Man, the Brute Creation, Vegetables, Minerals, Etc*. London, UK: Thomas Kelly, 1828.
———. *The Natural History of Quadrupeds*. Edinburgh, Scotland: Thomas

Nelson and Peter Brown, 1830.
Lidicker, William Z. (ed.). *Landscape Approaches in Mammalian Ecology and Conservation*. Minneapolis, MN: University of Minnesota Press, 1995.
Lindsay, Everett H., Volker Fahlbusch, and Pierre Mein (eds.). *European Neogene Mammal Chronology*. New York: Plenum Press, 1989.
Low, Kim Cheng Patrick. *Leading Successfully in Asia*. Cham, Switzerland: Springer, 2018.
Lydekker, Richard (ed.). *The Royal Natural History*. London, UK: Frederick Ware and Co., 1893-1894.
——. *Wildlife of the World: A Descriptive Survey of the Geographical Distribution of Animals*. London, UK: Frederick Warne, 1915.
Lydekker, Richard, et al. *Natural History*. New York: D. Appleton and Co., 1897.
——. *Guide to the Galleries of Mammals in the Department of Zoology of the British Museum (Natural History)*. London, UK: British Museum of Natural History, 1906.
Macdonald, David W. (ed.). *The Encyclopedia of Mammals*. 1995. Oxford, UK: Oxford University Press, 2006 ed.
——. *The Princeton Encyclopedia of Mammals*. Princeton, NJ: Princeton University Press, 2009.
Macdonald, David W., and Katrina Service (eds.). *Key Topics in Conservation Biology*. Malden, MA: Blackwell Publishing, 2007.
Macdonald, David W., and Andrew J. Loveridge (eds.). *Biology and Conservation of Wild Felids*. Oxford, UK: Oxford University Press, 2010.
MacMahon, James A. *The Audubon Society Nature Guides: Deserts*. New York: Alfred A. Knopf, 1985.
MacPhee, Ross D. E. *End of the Megafauna: The Fate of the World's Hugest, Fiercest, and Strangest Animals*. New York: W. W. Norton, 2018.
Mammal Anatomy: An Illustrated Guide. Tarrytown, NY: Marshall Cavendish Corp., 2010.
Mangin, Arthur. *The Desert World*. London, UK: T. Nelson and Sons, 1872.
Marks, Robert B. *Tigers, Rice, Silk, and Silt: Environment and Economy in Late Imperial South China*. Cambridge, UK: Cambridge University Press, 2004.
Martin, Paul S., and Richard G. Klein (eds.). *Quaternary Extinctions: A Prehistoric Revolution*. Tucson, AZ: University of Arizona Press, 1984.
Martin, Robert E., Ronald H. Pine, and Anthony F. DeBlase. *A Manual of*

Mammalogy: With Keys to Families of the World. 1974. Long Grove, IL: Waveland Press, 2011 ed.

Martyn, Thomas. *Elements of Natural History*. Cambridge, UK: J. Archdeacon, printer to the university, 1775.

Matthiessen, Peter. *Tigers in the Snow*. New York: North Point Press, 2000.

Maunder, Samuel. *The Treasury of Natural History: Or a Popular Dictionary of Animated Nature*. London,. UK: Longman, Brown, Green, and Longmans, 1852.

Mayer, William V., and Richard G. Van Gelder (eds.). *Physiological Mammalogy*. New York: Academic Press, 1963.

Mayr, Ernst. *Systematics and the Origin of Species from the Viewpoint of a Zoologist*. Cambridge, MA: Harvard University Press, 1999.

McDougal, Charles. *The Face of the Tiger*. London, UK: Rivington Books, 1977.

McKenna, Malcolm C., and Susan K. Bell. *Classification of Mammals: Above the Species Level*. New York: Columbia University Press, 1997.

McNamee, Thomas. *The Inner Life of Cats: The Science and Secrets of Our Mysterious Feline Companions*. Brentwood, TN: Hachette Books, 2018.

Meffe, Gary K., and C. Ronald Carroll. *Principles of Conservation Biology*. Sunderland, MA: Sinauer Associates, 1997.

Metford, J.C.J. *Dictionary of Christian Lore and Legend*. London, UK: Thames and Hudson, 1983.

Meyerhof, Wolfgang, and Sigrun Korsching (eds.). *Chemosensory Systems in Mammals, Fishes, and Insects*. Berlin, Germany: Springer, 2009.

Mieden, Wolfgang. *Proverbs: A Handbook*. Westport, CT: Greenwood Press, 2004.

Mish Frederick (ed.). *Webster's Ninth New Collegiate Dictionary*. 1828. Springfield, MA: Merriam-Webster, 1984 ed.

Mishra, Hemanta (with Jim Ottaway Jr.). *Bones of the Tiger: Protecting the Man-eaters of Nepal*. Guilford, CT: Lyons Press, 2010.

Mitra, Sudipta. *Gir Forest and the Saga of the Asiatic Lion*. New Delhi, India: Indus Publishing Co., 2005.

Monaghan, Patricia. *The Book of Goddesses and Heroines*. St. Paul, MN: Llewellyn Publications, 1990.

Morgan, Ben. *Guide to Mammals: A Wild Journey With These Extraordinary Beasts*. London, UK: DK Publishing, 2003.

Morrell, G. Herbert. *Comparative Anatomy, and Guide to Discussion, Part 1:*

Mammalia (Anatomy and Dissection). London, UK: Longman and Co., 1872.

Mukerji, Dhan Gopal. *Hari, the Jungle Lad.* New York: E. P. Dutton and Co., 1924.

Murray, Andrew. *The Geographical Distribution of Mammals.* London, UK: Day and Son, 1866.

Nagle, John, J. B. Ruhl, and Kalyani Robbins. *The Law of Biodiversity and Ecosystem Management.* St. Paul, MN: Foundation Press, 2012.

Nolan, Edward Henry. *The Illustrated History of the British Empire in India and the East, From the Earliest Times to the Suppression of the Sepoy Mutiny in 1859.* 2 vols. London, UK: James S. Virtue, 1857.

Nowak, Ronald M. *Walker's Mammals of the World.* 2 vols. 1964. Baltimore, MD: Johns Hopkins University Press, 1999 ed.

O'Brien, Stephen J., Joan C. Menninger, and William G. Nash (eds.). *Atlas of Mammalian Chromosomes.* Hoboken, NJ: John Wiley and Sons, 2006.

Owen, Richard. *On the Anatomy of Vertebrates, Vol. 3: Mammals.* London, UK: Longmans, Green, and Co., 1868.

Padel, Ruth. *Tigers in Red Weather: A Quest for the Last Wild Tigers.* New York: Walker and Co., 2009.

Page, Roderic D. M. (ed.). *Tangled Trees: Phylogeny, Cospeciation, and Coevolution.* Chicago, IL: University of Chicago Press, 2003.

Palmer, Ephraim Lawrence. *Fieldbook of Natural History.* New York: Whittlesey House, 1949.

Pennant, Thomas. *Synopsis of Quadrupeds.* Chester, UK: self-published, 1771.

Pitman, Allan. *100 Bedtime Stories for Triathletes.* Bloomington, IN: Balboa Press, 2017.

Prothero, Donald R. *After the Dinosaurs: The Age of Mammals.* Bloomington, IN: Indiana University Press, 2006.

———. *The Princeton Field Guide to Prehistoric Mammals.* Princeton, NJ: Princeton University Press, 2017.

Ratcliffe, Susan (ed.). *Oxford Treasury of Sayings and Quotations.* 1997. Oxford, UK: Oxford University Press, 2011 ed.

Rees, Paul A. *An Introduction to Zoo Biology and Management.* Oxford, UK: Wiley-Blackwell, 2011.

Reynolds, Sidney H. *British Pleistocene Mammalia.* London, UK: Paleontographical Society, 1906.

Rice, William. *Tiger-Shooting in India; Being an Account of Hunting Experiences on Foot in Rajpootana, During the Hot Seasons, From 1850 to 1854*. London, UK: Smith, Elder and Co., 1857.

Roest, Aryan I. *Key-Guide to Mammal Skulls and Lower Jaws*. Eureka, CA: Mad River Press, 1986.

Roosevelt, Theodore. *The Wilderness Hunter*. New York: G. P. Putnam's Sons, 1893.

Rose, Kenneth D. *The Beginning of the Age of Mammals*. Baltimore, MD: Johns Hopkins University Press, 2006.

Rose, Kenneth D., and J. David Archibald (eds.). *The Rise of Placental Mammals: Origins and Relationships of the Major Extant Clades*. Baltimore, MD: Johns Hopkins University Press, 2005.

Ryan, James M. *Mammalogy Techniques: Lab Manual*. Baltimore, MD: Johns Hopkins University Press, 2018.

Sadleir, Richard M. F. S. *The Ecology of Reproduction in Wild and Domestic Mammals*. 1969. New York: Springer, 2012 ed.

Sanderson, James G., and Patrick Wilson. *Small Wild Cats: The Animal Answer Guide*. Baltimore, MD: Johns Hopkins University Press, 2011.

Savage, R. J. G., and M. R. Long. *Mammal Evolution: A Brief Guide*. London, UK: Natural History Museum Publications, 1986.

Sawhney, Clifford (ed.). *The Book of Common and Uncommon Proverbs*. New Delhi, India: Pustak Mahal, 2008.

Schaller, George. *The Deer and the Tiger*. Chicago, IL: University Press of Chicago, 1967.

Schouler, James. *History of the United States of America, Under the Constitution*. Washington, D.C.: W. H. and O. H. Morrison, 1880.

Sclater, William Lutley, and Philip Lutley Sclater. *The Geography of Mammals*. London, UK: Kegan Paul, Trench, Trübner, and Co., 1899.

Seabrook, Lochlainn. *Christmas Before Christianity: How the Birthday of the "Sun" Became the Birthday of the "Son."* Franklin, TN: Sea Raven Press, 2010.

——. *Seabrook's Bible Dictionary of Traditional and Mystical Christian Doctrines*. Spring Hill, TN: Sea Raven Press, 2016.

——. *The Concise Book of Owls: A Guide to Nature's Most Mysterious Birds*. Spring Hill, TN: Sea Raven Press, 2019.

——. *North America's Amazing Mammals: An Encyclopedia for the Whole Family*. Spring Hill, TN: Sea Raven Press, 2020.

Seidensticker, John. *Great Cats: Majestic Creatures of the Wild*. Sydney,

Australia: Murdoch Books, 1991.

Seidensticker, John, Sarah Christie, and Peter Jackson (eds.). *Riding the Tiger: Tiger Conservation in Human-Dominated Landscapes.* Cambridge, UK: Cambridge University Press, 1999.

Server, Lee. *Tigers: A Portrait of the Animal World.* New York: New Line Books, 1998.

Seton, Ernest Thompson. *Wild Animals I Have Known.* New York: Grosset and Dunlap, 1898.

Shakespeare, William. *The Plays of William Shakespeare.* London, UK: J. Nichols and Son, 1813.

Shaw, Simeon. *Nature Displayed in the Heavens, and on the Earth, According to the Latest Discoveries.* 6 vols. London, UK: G. and W. B. Whitaker, 1823.

Shoberl, Frederic. *Natural History of Quadrupeds.* London, UK: John Harris, 1834.

Solway, Andrew. *Killer Cats.* Chicago, IL: Heinemann Library, 2005.

Soviet Life (January 1971). Washington, D.C.: Jointly published by the governments of the USA and USSR, 1971.

Stark, John. *Elements of Natural History, Adapted to the Present State of Science.* Edinburgh, Scotland: Adam Black and John Stark, 1828.

Stephens, James Francis. *General Zoology.* London, UK: self-published, 1826.

Sterndale, Robert Armitage. *Natural History of the Mammalia of India and Ceylon.* Calcutta, India: Thacker, Spink, and Co., 1884.

Stidworthy, John. *Mammals: The Large Plant-Eaters.* New York: Facts on File, 1989.

Stone, Julia A. *Illustrated India: Its Princes and People.* Hartford, CT: American Publishing Co., 1877.

Strong, Asa B. (ed.). *Illustrated Natural History of the Three Kingdoms.* New York: Green and Spencer, 1849.

Sunquist, Melvin E., and Fiona Sunquist. *Wild Cats of the World.* Chicago, IL: University of Chicago Press, 2002.

Szalay, Frederick S., Michael J. Novacek, and Malcolm C. McKenna (eds.). *Mammal Phylogeny: Mesozoic Differentiation, Multituberculates, Monotremes, Early Therians, and Marsupials.* New York: Springer-Verlag, 1993.

Thapur, Valmik (ed.). *Saving Wild Tigers, 1900-2000: The Essential Writings.* Delhi, India: Permanent Black, 2001.

———. *Wild Tigers of Ranthambhore*. Oxford, UK: Oxford University Press, 2005.

The Columbia Encyclopedia. New York: Columbia University Press, 1964.

The English Encyclopaedia. London, UK: G. Kearsley, 1802.

The English Illustrated Magazine (1891-1892). London, UK: Macmillan and Co., 1892.

The Hutchinson Encyclopedia. Oxford, UK: Helicon Publishing, 2001.

The London Encyclopedia. 22 vols. London, UK: Thomas Tegg, 1845.

The New American Encyclopedia. New York: D. Appleton and Co., 1862.

The New Encyclopedia Britannica. 1768. London, UK: Encyclopedia Britannica, Inc., 1975 ed.

The New Larousse Encyclopedia of Animal Life. 1967. New York: Bonanza Books, 1980 ed.

The Oracle Encyclopedia. London, UK: George Newnes, 1896.

The Sporting Almanac, 1841. London, UK: n.p., 1841.

Tilson, Ronald, and Philip J. Nyhus (eds.). *Tigers of the World: The Science, Politics, and Conservation of Panthera Tigris*. 1987. London, UK: Academic Press, 2010 ed.

Townsend, Colin R., Michael Begon, and John L. Harper. *Essentials of Ecology* (2nd ed.). 2000. Malden, MA: Blackwell Science, 2003 ed.

Trimmer, Mary. *A Natural History of the Most Remarkable Quadrupeds, Birds, Fishes, Serpents, Reptiles, and Insects*. Chiswick, UK: self-published, 1825.

Turner, Dennis C., and Patrick Bateson (eds.). *The Domestic Cat: The Biology of its Behaviour*. 1988. Cambridge, UK: Cambridge University Press, 2014 ed.

Ungar, Peter S. *Mammal Teeth: Origin, Evolution, and Diversity*. Baltimore, MD: Johns Hopkins University Press, 2010.

United States Department of Agriculture. *Our Forests: What They Are and What They Mean to Us*. Washington, D.C.: U.S. Government Printing Office, 1950.

United States Department of the Army. *Area Handbook for Korea*. Washington, D.C.: U.S. Government Printing Office, 1964.

United States Department of the Interior. *Tiger: Panthera Tigris*. Washington, D.C.: U.S. Government Printing Office, 1994.

———. *Rhinoceros and Tiger Conservation Act: Summary Report 1996-1998*. Washington, D.C.: U.S. Government Printing Office, 1999.

United States Subcommittee on Fisheries, Conservation, Wildlife and

Oceans. *Rhino and Tiger Conservation*. Washington, D.C.: U.S. Government Printing Office, 1998.

USSR Illustrated Monthly (January 1964). Washington, D.C.: Jointly published by the governments of the USA and USSR, 1964.

Vaillant, John. *The Tiger: A True Story of Vengeance and Survival*. New York: Random House, 2010.

Vaughan, Terry A., James M. Ryan, and Nicholas J. Czaplewski. *Mammalogy*. Sudbury, MA: Jones and Bartlett, 2011.

Verne, Jules. *The Steam House*. New York: Charles Scribner's Sons, 1881.

Vitullo-Martin, Julia, and J. Robert Moskin. *The Executive's Book of Quotations: A Guide to the Right Quote for Every Occasion*. New York: Oxford University Press, 1994.

Vojnich, Oszkár. *In the East Indian Archipelago*. Budapest, Hungary: Singer and Wolfner, 1913.

Walker, Barbara G. *The Woman's Dictionary of Symbols and Sacred Objects*. San Francisco, CA: Harper and Row, 1988.

Walker, Ernest Pillsbury. *Mammals of the World*. 2 vols. 1968. Baltimore, MD: Johns Hopkins University Press, 1975 ed.

Ward, William. *A View of the History, Literature, and Mythology of the Hindoos*. 3 vols. London, UK: self-published, 1822.

Weigl, Richard. *Longevity of Animals in Captivity: From the Living Collections of the World*. Stuttgart, Germany: Kleine Senckenberg-Reihe, 2005.

Wheeler, J. Talboys. *The History of India From the Earliest Ages*. London, UK: N. Trübner and Co., 1869.

Whitney, William Dwight. *The Century Dictionary: An Encyclopedic Lexicon of the English Language*. 1889. New York: The Century Co., 1911 ed.

Whyte, Adam Gowans. *The Wonder World We Live In*. New York: Alfred A. Knopf, 1921.

Wilson, Don E., and DeeAnn M. Reeder (eds.). *Mammal Species of the World: A Taxonomic and Geographic Reference*. 2 vols. Baltimore MD: Johns Hopkins University Press, 2005.

Wilson, Don E., and F. Russell Cole. *Common Names of Mammals of the World*. Washington, D.C.: Smithsonian Institution Scholarly Press, 2000.

Wilson, Edward O. *Sociobiology: The New Synthesis*. Cambridge, MA: Belknap Press, 1975.

Winter, Steve, with Sharon Guynup. *Tigers Forever: Saving the World's Most Endangered Big Cat*. Washington, D.C.: National Geographic, 2013.

Wood, John George. *The Illustrated Natural History*. London, UK: George Routledge and Sons, 1853.

——. *Natural History Picture Book for Children*. London, UK: Routledge, Warne, and Routledge, 1861.

Woodburn, Michael O. *Late Cretaceous and Cenozoic Mammals of North America: Biostratigraphy and Geochronology*. New York: Columbia University Press, 2004.

World Checklist of Threatened Mammals. Peterborough, UK: Joint Nature Conservation Committee, 1993.

Wright, John (ed.). *A Natural History of the Globe, of Man, of Beasts, Birds, Fishes, Reptiles, Insects, and Plants*. 5 vols. Philadelphia, PA: Thomas DeSilver Jr., 1831.

Wright, Thomas (ed.). *The William Blake Calendar: A Quotation From the Works of William Blake for Every Day in the Year*. London, UK: Frank Palmer, 1913.

Wrobel, Murray (ed.). *Elsevier's Dictionary of Mammals*. Amsterdam, The Netherlands, Elsevier, 2007.

Young, Stanley Paul, and E. A. Goldman. *The Puma: Mysterious American Cat*. Washington, D.C.: American Wildlife Institute, 1946.

Zachos, Frank E. *Species Concepts in Biology: Historical Development, Theoretical Foundations and Practical Relevance*. Basel, Switzerland: Springer, 2016.

Zachos, Frank E., and Robert J. Asher (eds.). *Mammalian Evolution, Diversity and Systematics*. Berlin, Germany: De Gruyter, 2018.

Ziaie, Hooshang. *A Field Guide to the Mammals of Iran*. Tehran, Iran: Dept. of Environment, 1996.

A tiger hunt gone awry.

INDEX

- ONLY FELINES ARE LISTED IN THE INDEX.
- LOOK UP FELINES BY THEIR COMMON NAMES, ALTERNATE NAMES, OR SCIENTIFIC NAMES.
- FOR EASIER READING SCIENTIFIC NAMES ARE NOT ITALICIZED.

Acinonyx jubatus, 39
African golden cat, 39
African wild cat, 39
Amoy tiger, 116
Amur tiger, 19, 22, 42-44, 99-101, 104, 107
Andean cat, 39
Antique tiger, 122
Asiatic golden cat, 39
Bali tiger, 22, 42, 102, 103
Balinese tiger, 102
Bengal tiger, 19, 25, 26, 34, 43, 44, 104-106, 114
black tiger, 26, 104
black-footed cat, 39
blue tiger, 26
bobcat, 39
Borneo bay cat, 39
Canada lynx, 39
caracal, 39
Caracal aurata, 39
Caracal caracal, 39
Caspian tiger, 43, 101, 106, 107
Catopuma badia, 39
Catopuma temminckii, 39
cheetah, 39
Chinese mountain cat, 39
clouded leopard, 39, 40
Corbett's tiger, 108
domestic cat, 33, 34, 39
Eurasian lynx, 39
European wildcat, 39
Felis bieti, 39
Felis catus, 33, 39
Felis chaus, 39
Felis lybica, 39
Felis margarita, 39
Felis nigripes, 39

Felis silvestris, 39
fishing cat, 39
flat-headed catfish, 39
Geoffroy's cat, 39
guiña, 39
Harimau Belang, 112
Herpailurus yagouaroundi, 39
Hyrcanian tiger, 106
Iberian lynx, 39
Indian tiger, 44, 104, 105
Indochinese tiger, 22, 43, 47, 108, 109, 113
jaguar, 15, 39, 40, 45, 62
jaguarundi, 39
Javan tiger, 43, 103, 110, 111
jungle cat, 39
leopard, 15, 39, 40, 45, 67, 83, 100
Leopardus colocolo, 39
Leopardus geoffroyi, 39
Leopardus guigna, 39
Leopardus guttulus, 40
Leopardus jacobita, 39
Leopardus pardalis, 39
Leopardus tigrinus, 39
Leopardus wiedii, 39
Leptailurus serval, 39
liger, 17
lion, 15, 17, 39, 40, 45, 51, 59, 62
Lynx canadensis, 39
Lynx lynx, 39
Lynx pardinus, 39
Lynx rufus, 39
mainland leopard cat, 39
Malayan tiger, 43, 108, 109, 112
Malayan tigers, 113
Maltese tiger, 26
Manchurian tiger, 99
marbled cat, 39

margay, 39
Mazandaran tiger, 106
mountain lion, 39
Neofelis diardi, 40
Neofelis nebulosa, 39, 40
Ngandong tiger, 43, 114, 115
Northeast China tiger, 99
northern oncilla, 39
ocelot, 39
Otocolobus manul, 39
Pallas' cat, 39
Pampas cat, 39
Panthera leo, 39, 40, 45
Panthera onca, 39, 40, 45
Panthera pardus, 39, 40, 45
Panthera tigris, 18, 21, 33, 37, 38, 40, 43, 44, 48, 54, 61, 67, 68, 70, 71, 76, 82, 93, 95, 106, 114
Panthera tigris acutidens, 43, 122
Panthera tigris altaica, 11, 42-44, 99
Panthera tigris amoyensis, 43, 116
Panthera tigris balica, 42, 102
Panthera tigris bengalensis, 104
Panthera tigris corbetti, 43, 108, 109, 113
Panthera tigris jacksoni, 43, 108, 109, 112, 113
Panthera tigris malayensis, 112
Panthera tigris soloensis, 43, 114
Panthera tigris sondaica, 43, 110
Panthera tigris sumatrae, 43, 118
Panthera tigris tigris, 25, 43, 44, 104
Panthera tigris trinilensis, 43, 120
Panthera tigris virgata, 43, 106
Panthera uncia, 40, 45
Pardofelis marmorata, 39
Persian tiger, 106
Pleistocene tiger, 120
Prionailurus bengalensis, 39
Prionailurus javanensis, 40
Prionailurus planiceps, 39
Prionailurus rubiginosus, 39
Prionailurus viverrinus, 39
Proailurus, 46

Pseudaelurus, 47
Puma concolor, 39
road leopard, 106
royal Bengal tiger, 45, 82, 104
rusty-spotted cat, 39
sand cat, 39
Schizailurus, 47
serval, 39
Siberian tiger, 43, 99, 100
snow leopard, 40, 45, 46
South China tiger, 26, 43, 48, 116, 117, 122
southern oncilla, 40
Sumatran tiger, 43, 83, 110, 118, 119
Sunda clouded leopard, 40
Sunda leopard cat, 40
tigon, 17
traveling leopard, 106
Trinil tiger, 43, 120, 121
Turan tiger, 106
Turanian tiger, 106
Ussurian, tiger, 99
Wanhsien tiger, 43, 122
white Bengal tiger, 45
white tiger, 25, 72, 75, 104
Xiamen tiger, 116

One of the world's most dangerous and merciless creatures: a wounded tiger.

International Tiger Day

Remember July 29, International Tiger Day, held every year on this day in an effort to bring increased awareness to the plight of tigers worldwide.

Panthera tigris.

MEET THE AUTHOR

LOCHLAINN SEABROOK is a lifelong writer, naturalist, and award-winning author of books ranging in topic from nature and science to history and religion. He wrote his first natural history composition at age nine, an essay on the fisher (*Martes pennanti*).

His love of animals, natural history, and the great outdoors has taken him to nearly every corner of the U.S., from working as a hunting guide in the Rocky Mountains to ranch hand on the Great Plains, from farrier assistant in the Northeast to stableman in the Southwest, from wrangler in the Midwest to fish farmer in the Deep South. One of his most memorable experiences was helping deliver a foal in a barn at midnight on a freezing spring evening as a snowstorm raged outside.

An eagle scout, a Kentucky Colonel, a nature photographer, and a 17th-generation Southerner of Appalachian heritage, he also worked as a zookeeper and animal handler at a private wildlife park and sanctuary, where he cared for the following animals: birds of prey (owl, hawk, eagle, vulture); reptiles (alligator, snake, turtle); small mammals (mouse, rat, chipmunk); medium-sized mammals (rabbit, skunk, prairie dog, otter); large mammals (bear, deer, horse); canids (fox, coyote); medium-sized cats (bobcat, lynx, margay); large cats (Amur tiger, mountain lion, jaguar).

In addition to *The Concise Book of Tigers: A Guide to Nature's Most Remarkable Cats*, Colonel Seabrook is also the author of the bestseller, *The Concise Book of Owls: A Guide to Nature's Most Mysterious Birds* (endorsed by the World Bird Sanctuary), as well as the monumental work, *North America's Amazing Mammals: An Encyclopedia for the Whole Family* (the world's first scientific animal guide to include an entry on Sasquatch).

For more info visit

LochlainnSeabrook.com

God made the cat to give Man the pleasure of stroking a tiger.

François-Joseph Méry
(1798-1866)

If you enjoyed this book you will be interested in Colonel Seabrook's related titles:

☞ THE CONCISE BOOK OF OWLS: A GUIDE TO NATURE'S MOST MYSTERIOUS BIRDS
☞ NORTH AMERICA'S AMAZING MAMMALS: AN ENCYCLOPEDIA FOR THE WHOLE FAMILY

Available from Sea Raven Press and wherever fine books are sold

ALL OF OUR BOOK COVERS ARE AVAILABLE AS 11" X 17" POSTERS, SUITABLE FOR FRAMING

SeaRavenPress.com

www.ingramcontent.com/pod-product-compliance
Lightning Source LLC
Chambersburg PA
CBHW062111080426
42734CB00012B/2826